CliffsNotes®
Algebra I
Quick Review

By Jerry Bobrow, Ph.D.
Revised by Ed Kohn, M.S.
2nd Edition

D0963838

CliffsNotes®
Algebra I
Quick Review

By Jerry Bobrow, Ph.D.
Revised by Ed Kohn, M.S.
2nd Edition

Houghton Mifflin Harcourt
Boston • New York

About the Author

Jerry Bobrow, Ph.D., was an award-winning teacher and educator, and his company Bobrow Test Preparation Services is a national authority in the field of test preparation. Bobrow Test Preparation Services has been administering test preparation programs for most California State Universities for the past 34 years. Dr. Bobrow and his faculty have authored more than 30 national best-selling test preparation books including Cliffs Preparation Guides for the GRE, GMAT, CSET, SAT, CBEST, NTE, ACT, and PPST. Each year the faculty of Bobrow Test Preparation Services lectures to thousands of students on preparing for these important exams.

Publisher's Acknowledgments

Editorial

Acquisitions Editor: Greg Tubach
Project Editor: Kelly D. Henthorne
Technical Editors: David Herzog and
    Mary Jane Sterling

Composition

Indexer: Potomac Indexing, LLC
Proofreader: Henry Lazarek
Wiley Publishing, Inc. Composition Services
Consultant Ron Podrasky, M.A.

Library of Congress Cataloging-in-Publication data is available.
ISBN 978-0-470-88028-9 (pbk)

Printed in the United States of America
10 9 8 7 6 5

Table of Contents

INTRODUCTION

CliffsNotes Algebra I Quick Review is designed to give a clear, concise, easy-to-use review of the basics of algebra. Introducing each topic, defining key terms, and carefully *walking* through each sample problem type in a step-by-step manner gives the student insight and understanding to important algebraic concepts.

The prerequisite to get the most out of this book is an understanding of the important concepts of basic math—working with fractions, decimals, percents, and signed numbers. *CliffsNotes Algebra I Quick Review* starts with a short review of pre-algebra (Chapters 1 and 2) to review some necessary background. The rest of the book (Chapters 3 through 13) focuses on the foundations of algebra.

Why You Need This Book

Can you answer "yes" to any of these questions?

- Do you need to review the fundamentals of algebra?
- Do you need a course supplement to Algebra I?
- Do you need a concise, comprehensive reference for algebraic concepts?

If so, then *CliffsNotes Algebra I Quick Review* is for you!

How to Use This Book

You can use this book in any way that fits your personal style for study and review—you decide what works best with your needs. You can either read the book from cover to cover or just look for the information you want and put it back on the shelf for later. Here are just a few ways you can search for topics:

- Look for areas of interest in the book's table of contents or use the index to find specific topics.

- Flip through the book, looking for subject areas at the top of each page.

- Get a glimpse of what you'll gain from a chapter by reading through the "Chapter Check-In" at the beginning of each chapter.

- Use the "Chapter Check-Out" at the end of each chapter to gauge your grasp of the important information you need to know.

- Test your knowledge more completely in the "Review Questions" and look for additional sources of information in the "Resource Center."

- Use the glossary to find key terms fast. This book defines new terms and concepts where they first appear in the chapter. If a word is boldfaced, you can find a more complete definition in the book's glossary.

- Or flip through the book until you find what you're looking for— we organized this book to gradually build on key concepts.

Hundreds of Practice Questions Online!

Prepare for your next Algebra I quiz or test with hundreds of additional practice questions online. The questions are organized by this book's chapter sections, so it's easy to use the book and then quiz yourself online to make sure you know the subject. Go to www.cliffsnotes.com/go/quiz/algebra_i to test yourself anytime and find other free homework help.

CHAPTER 1

PRELIMINARIES AND BASIC OPERATIONS

Chapter Check-In

❑ Categories of numbers

❑ Properties of addition and multiplication

❑ Powers and exponents

❑ Square roots and cube roots

❑ Parentheses, brackets, and braces

❑ Divisibility rules

Before you begin learning, relearning, or reviewing algebra, you need to feel comfortable with some pre-algebra terms and operations. This chapter starts with some basic essentials.

Preliminaries

The first items you should become familiar with are the different categories or types of numbers and the common math symbols.

Categories of numbers

In doing algebra, you work with several categories of numbers.

- **Natural or counting numbers.** The numbers 1, 2, 3, 4, . . . are called *natural* or *counting numbers.*

- **Whole numbers.** The numbers 0, 1, 2, 3, . . . are called *whole numbers.*

- **Integers.** The numbers . . . –2, –1, 0, 1, 2, . . . are called *integers.*

- **Negative integers.** The numbers . . . –3, –2, –1 are called negative *integers*.

- **Positive integers.** The natural numbers are sometimes called the *positive integers*.

- **Rational numbers.** Fractions, such as $\frac{3}{2}$ or $\frac{7}{8}$, are called *rational numbers*. Since a number such as 5 may be written as $\frac{5}{1}$, all *integers* are *rational numbers*. All rational numbers can be written as fractions $\frac{a}{b}$, with a being an integer and b being a natural number. Terminating and repeating decimals are also rational numbers, because they can be written as fractions in this form.

- **Irrational numbers.** Another type of number is an *irrational number*. Irrational numbers *cannot* be written as fractions $\frac{a}{b}$, with a being an integer and b being a natural number. $\sqrt{3}$ and π are examples of irrational numbers. *An irrational* number, when exactly expressed as a decimal, neither terminates nor has a repeating decimal pattern.

- **Even numbers.** *Even numbers* are integers divisible by 2: … –6, –4, –2, 0, 2, 4, 6, …

- **Prime numbers.** A *prime number* is a natural number that has exactly two different factors, or that can be perfectly divided by only itself and 1. For example, 19 is a prime number because it can be perfectly divided by only 19 and 1, but 21 is not a prime number because 21 can be perfectly divided by other numbers (3 and 7). The only even prime number is 2; thereafter, any even number may be divided perfectly by 2. Zero and 1 are not prime numbers or composite numbers. The first 10 prime numbers are 2, 3, 5, 7, 11, 13, 17, 19, 23, and 29.

- **Odd numbers.** *Odd numbers* are integers not divisible by 2: …–5, –3, –1, 1, 3, 5,….

- **Composite numbers.** A *composite number* is a natural number divisible by more than just 1 and itself: …4, 6, 8, 9,…

- **Squares.** *Squares* are the result when numbers are multiplied by themselves, that is, raised to the second power. $2 \cdot 2 = 4$; $3 \cdot 3 = 9$. The first six squares of natural numbers are 1, 4, 9, 16, 25, 36.

- **Cubes.** *Cubes* are the result when numbers are multiplied by themselves and then again by the original number, that is, raised to the third power. $2 \cdot 2 \cdot 2 = 8$; $3 \cdot 3 \cdot 3 = 27$. The first six cubes of natural numbers are 1, 8, 27, 64, 125, 216.

Ways to show multiplication

There are several ways to show multiplication of a pair of numerical values.

■ When the two numerical values are known, you can show the multiplication of 4 with 3 as follows:

$$4 \times 3$$

$$4 \cdot 3$$

$$(4)(3)$$

$$4(3)$$

$$(4)3$$

■ When one value is a number and the other value is a variable: show the multiplication of 4 and a as follows:

$$4 \times a$$

$$4 \cdot a$$

$$(4)(a)$$

$$4(a)$$

$$(4)a$$

$$4a$$

■ When both values are variables: show the multiplication of a and b as follows:

$$a \times b$$

$$a \cdot b$$

$$(a)(b)$$

$$a(b)$$

$$(a)b$$

$$ab$$

Common math symbols

The following math symbols appear throughout algebra. Be sure to know what each symbol represents.

Symbol references:

= is equal to

≠ is not equal to

> is greater than

< is less than

≥ is greater than or equal to (also written ≧)

≤ is less than or equal to (also written ≦)

$\not>$ is not greater than

$\not<$ is not less than

$\not\geq$ is not greater than or equal to

$\not\leq$ is not less than or equal to

≈ is approximately equal to (also ≐)

Properties of Basic Mathematical Operations

Some mathematical operations have properties that can make them easier to work with and can actually save you time.

Some properties (axioms) of addition

You should know the definition of each of the following properties of addition and how each can be used.

- **Closure** is when all answers fall into the original set. If you add two even numbers, the answer is still an even number (2 + 4 = 6); therefore, the set of even numbers *is closed* under addition (has closure). If you add two odd numbers, the answer is not an odd number (3 + 5 = 8); therefore, the set of odd numbers is *not closed* under addition (no closure).

- **Commutative** means that the *order* does not make any difference in the result.

$$2 + 3 = 3 + 2$$

$$a + b = b + a$$

Note: Commutative does not hold for subtraction.

$$3 - 1 \neq 1 - 3$$

$$2 \neq -2$$

$$a - b \neq b - a$$

- **Associative** means that the *grouping* does not make any difference in the result.

$$(2 + 3) + 4 = 2 + (3 + 4)$$

$$(a + b) + c = a + (b + c)$$

The grouping has changed (parentheses moved), but the sides are still equal.

Note: Associative does *not* hold for subtraction.

$$4 - (3 - 1) \neq (4 - 3) - 1$$

$$4 - 2 \neq 1 - 1$$

$$2 \neq 0$$

$$a - (b - c) \neq (a - b) - c$$

- The **identity element** for addition is 0. Any number added to 0 gives the original number.

$$3 + 0 = 0 + 3 = 3$$

$$a + 0 = 0 + a = a$$

- The **additive inverse** is the opposite (negative) of the number. Any number plus its additive inverse equals 0 (the identity).

$3 + (-3) = 0$; therefore, 3 and -3 are additive inverses.

$-2 + 2 = 0$; therefore, -2 and 2 are additive inverses.

$a + (-a) = 0$; therefore, a and $-a$ are additive inverses.

Some properties (axioms) of multiplication

You should know the definition of each of the following properties of multiplication and how each can be used.

- **Closure** is when all answers fall into the original set. If you multiply two even numbers, the answer is still an even number ($2 \times 4 = 8$); therefore, the set of even numbers *is closed* under multiplication (has closure). If you multiply two odd numbers, the answer is an odd number ($3 \times 5 = 15$); therefore, the set of odd numbers *is closed* under multiplication (has closure).

- **Commutative** means the *order* does not make any difference in the result.

$$2 \times 3 = 3 \times 2$$

$$a \times b = b \times a$$

Note: Commutative does *not* hold for division.

$$2 \div 4 \neq 4 \div 2$$

$$\frac{2}{4} \neq \frac{4}{2}$$

$$\frac{1}{2} \neq 2$$

$$a \div b \neq b \div a$$

■ **Associative** means that the *grouping* does not make any difference in the result.

$$(2 \times 3) \times 4 = 2 \times (3 \times 4)$$

$$(a \times b) \times c = a \times (b \times c)$$

The grouping has changed (parentheses moved) but the sides are still equal.

Note: Associative does *not* hold for division.

$$(8 \div 4) \div 2 \neq 8 \div (4 \div 2)$$

$$2 \div 2 \neq 8 \div 2$$

$$1 \neq 4$$

$$(a \div b) \div c \neq a \div (b \div c)$$

■ The **identity element** for multiplication is 1. Any number multiplied by 1 gives the original number.

$$3 \times 1 = 1 \times 3 = 3$$

$$a \times 1 = 1 \times a = a$$

■ The **multiplicative inverse** is the **reciprocal** of the number. Any nonzero number multiplied by its reciprocal equals 1.

$2 \times \frac{1}{2} = 1$; therefore, 2 and $\frac{1}{2}$ are multiplicative inverses.

$a \times \frac{1}{a} = 1$; therefore, a and $\frac{1}{a}$ are multiplicative inverses

(provided $a \neq 0$).

A property of two operations

The distributive property is the process of passing the number value outside of the parentheses, using multiplication, to the numbers being added or subtracted inside the parentheses. In order to apply the distributive property, it must be multiplication outside the parentheses and either addition or subtraction inside the parentheses.

$$2(3 + 4) = 2(3) + 2(4) \qquad 5(12 - 3) = 5(12) - 5(3)$$
$$2(7) = 6 + 8 \qquad 5(9) = 60 - 15$$
$$14 = 14 \qquad 45 = 45$$
$$a(b + c) = a(b) + a(c) \qquad a(b - c) = a(b) - a(c)$$

Note: You cannot use the distributive property with only one operation.

$$3(4 \times 5 \times 6) \neq 3(4) \times 3(5) \times 3(6)$$
$$3(120) \neq 12 \times 15 \times 18$$
$$360 \neq 3240$$
$$a(bcd) \neq a(b) \times a(c) \times a(d) \text{ or}$$
$$a(bcd) \neq (ab)(ac)(ad)$$

Multiplying and Dividing Using Zero

Zero times any number equals zero.

$$0 \times 5 = 0$$
$$0 \times (-3) = 0$$
$$8 \times 9 \times 3 \times (-4) \times 0 = 0$$

Likewise, zero divided by any nonzero number is zero.

$$0 \div 5 = 0$$
$$0 \div (-6) = 0$$

Important note: Dividing by zero is "undefined" and is not permitted.

$\frac{6}{0}$ and $\frac{0}{0}$ are not permitted.

$\frac{6}{0}$ has no answer and $\frac{0}{0}$ does not have a unique answer.

In neither case is the answer zero.

Powers and Exponents

An **exponent** is a positive or negative number placed above and to the right of a quantity. It expresses the *power* to which the quantity is to be raised or lowered. In 4^3, 3 is the exponent and 4 is called the base. It shows that 4 is to be used as a factor three times. $4 \times 4 \times 4$ (multiplied by itself twice). 4^3 is read as *four to the third power* (or *four cubed* as discussed later in this chapter).

$$2^4 = 2 \times 2 \times 2 \times 2 = 16$$

$$3^2 = 3 \times 3 = 9$$

Remember that $x^1 = x$ and $x^0 = 1$ when x is any number (other than 0).

$2^1 = 2$	$2^0 = 1$
$3^1 = 3$	$3^0 = 1$
$4^1 = 4$	$4^0 = 1$

If the exponent is negative, such as 3^{-2}, then the base can be dropped under the number 1 in a fraction and the exponent made positive. An alternative method is to take the reciprocal of the base and change the exponent to a positive value.

Example 1: Simplify the following by changing the exponent from a negative value to a positive value and then evaluate the expression.

Under 1 Method   Reciprocal Method

(a) 3^{-2} $3^{-2} = \dfrac{1}{3^2} = \dfrac{1}{9}$ $3^{-2} = \left(\dfrac{1}{3}\right)^2 = \dfrac{1}{9}$

(b) 2^{-3} $2^{-3} = \dfrac{1}{2^3} = \dfrac{1}{8}$ $2^{-3} = \left(\dfrac{1}{2}\right)^3 = \dfrac{1}{8}$

(c) $\left(\dfrac{2}{3}\right)^{-4}$ $\left(\dfrac{2}{3}\right)^{-4} = \dfrac{1}{\left(\dfrac{2}{3}\right)^4} = \dfrac{1}{\left(\dfrac{16}{81}\right)} = \dfrac{81}{16}$ $\left(\dfrac{2}{3}\right)^{-4} = \left(\dfrac{3}{2}\right)^4 = \dfrac{81}{16}$

Squares and cubes

Two specific types of powers should be noted, **squares** and **cubes.** To *square a number,* just multiply it by itself (the exponent would be 2). For example, 6 squared (written 6^2) is 6×6, or 36. 36 is called a *perfect square* (the square of a whole number). Following is a list of the first twelve perfect squares:

$1^2 = 1$	$5^2 = 25$	$9^2 = 81$
$2^2 = 4$	$6^2 = 36$	$10^2 = 100$
$3^2 = 9$	$7^2 = 49$	$11^2 = 121$
$4^2 = 16$	$8^2 = 64$	$12^2 = 144$

To *cube a number*, just multiply it by itself twice (the exponent would be 3). For example, 5 cubed (written 5^3) is $5 \times 5 \times 5$, or 125. 125 is called a *perfect cube* (the cube of a whole number). Following is a list of the first twelve perfect cubes.

$1^3 = 1$	$5^3 = 125$	$9^3 = 729$
$2^3 = 8$	$6^3 = 216$	$10^3 = 1000$
$3^3 = 27$	$7^3 = 343$	$11^3 = 1331$
$4^3 = 64$	$8^3 = 512$	$12^3 = 1728$

Operations with powers and exponents

To *multiply* two numbers with exponents, *if the base numbers are the same,* simply keep the base number and add the exponents.

Example 2: Multiply the following, leaving the answers with exponents.

(a) $2^3 \times 2^5 = 2^8$ $(2 \times 2 \times 2) \times (2 \times 2 \times 2 \times 2 \times 2) = 2^8$
(b) $3^2 \times 3^4 = 3^6$ $(3 \times 3) \times (3 \times 3 \times 3 \times 3) = 3^6$

To *divide* two numbers with exponents, *if the base numbers are the same*, simply keep the base number and subtract the second exponent from the first, or the exponent of the **denominator** from the exponent of the **numerator**.

Example 3: Divide the following, leaving the answers with exponents.

(a) $4^8 \div 4^5 = 4^3$ $4^8 \div 4^5 = \dfrac{4^8}{4^5}$

$$\frac{4 \cdot 4 \cdot 4 \cdot 4 \cdot 4 \cdot 4 \cdot 4 \cdot 4}{4 \cdot 4 \cdot 4 \cdot 4 \cdot 4} = 4 \cdot 4 \cdot 4 = 4^3$$

(b) $\dfrac{9^6}{9^2} = 9^4$ $\dfrac{9^6}{9^2} = \dfrac{9 \cdot 9 \cdot 9 \cdot 9 \cdot 9 \cdot 9}{9 \cdot 9} = 9 \cdot 9 \cdot 9 \cdot 9 = 9^4$

To multiply or divide numbers with exponents, if the base numbers are different, you must simplify each number with an exponent first and then perform the operation.

Example 4: Simplify and perform the operation indicated.

(a) $3^2 \times 2^2 = 9 \times 4 = 36$
(b) $6^2 \div 2^3 = 36 \div 8 = 4\dfrac{4}{8} = 4\dfrac{1}{2}$

(Some shortcuts are possible.)

To *add* or *subtract* numbers with exponents, *whether the base numbers are the same or different,* you must simplify each number with an exponent first and then perform the indicated operation.

Example 5: Simplify and perform the operation indicated.

(a) $3^2 - 2^3 = 9 - 8 = 1$

(b) $4^3 + 3^2 = 64 + 9 = 73$

If a *number with an exponent is raised to another power* $(4^2)^3$, simply keep the original base number and multiply the exponents.

Example 6: Multiply and leave the answers with exponents.

(a) $(4^2)^3 = (4^2)(4^2)(4^2) = 4^6$ or $(4^2)^3 = 4^{(2)(3)} = 4^6$

(b) $(3^3)^2 = (3^3)(3^3) = 3^6$ or $(3^3)^2 = 3^{(3)(2)} = 3^6$

Square Roots and Cube Roots

Note that square and cube roots and operations with them are often included in algebra sections, and both topics are discussed more in Chapter 11.

Square roots. To find the **square root** of a number, you want to find some number that when multiplied by itself gives you the original number. In other words, to find the square root of 25, you want to find the number that when multiplied by itself gives you 25. The square root of 25, then, is 5. The symbol for square root is $\sqrt{}$. Following is a list of the first eleven perfect (whole number) square roots.

$$\sqrt{0} = 0 \qquad \sqrt{16} = 4 \qquad \sqrt{64} = 8$$
$$\sqrt{1} = 1 \qquad \sqrt{25} = 5 \qquad \sqrt{81} = 9$$
$$\sqrt{4} = 2 \qquad \sqrt{36} = 6 \qquad \sqrt{100} = 10$$
$$\sqrt{9} = 3 \qquad \sqrt{49} = 7$$

Special note: If no sign (or a positive sign) is placed in front of the square root, then the positive answer is required. Only if a negative sign is in front of the square root is the negative answer required. This notation is used in many texts and is adhered to in this book. Therefore,

$$\sqrt{9} = 3 \qquad \text{and} \qquad -\sqrt{9} = -3$$

Cube roots

To find the **cube root** of a number, you want to find some number that when multiplied by itself twice gives you the original number. In other words, to find the cube root of 8, you want to find the number that when multiplied by itself twice gives you 8. The cube root of 8, then, is 2, because $2 \times 2 \times 2 = 8$. Notice that the symbol for cube root is the radical sign with a small three (called the *index*) above and to the left $\sqrt[3]{}$. Other roots are defined similarly and identified by the index given. (In square root, an index of two is understood and usually not written.) Following is a list of the first eleven *perfect* (whole number) *cube roots*.

$\sqrt[3]{0} = 0$	$\sqrt[3]{64} = 4$	$\sqrt[3]{512} = 8$
$\sqrt[3]{1} = 1$	$\sqrt[3]{125} = 5$	$\sqrt[3]{729} = 9$
$\sqrt[3]{8} = 2$	$\sqrt[3]{216} = 6$	$\sqrt[3]{1000} = 10$
$\sqrt[3]{27} = 3$	$\sqrt[3]{343} = 7$	

Approximating square roots

To find the square root of a number that is not a perfect square, it will be necessary to find an *approximate* answer by using the procedure given in Example 7.

Example 7: Approximate $\sqrt{42}$.

Since $6^2 = 36$ and $7^2 = 49$, then $\sqrt{42}$ is between $\sqrt{36}$ and $\sqrt{49}$.

Therefore, $\sqrt{42}$ is a value between 6 and 7. Since 42 is about halfway between 36 and 49, you can expect that $\sqrt{42}$ will be close to halfway between 6 and 7, or about 6.5. To check this estimation, $6.5 \times 6.5 = 42.25$, or about 42.

Square roots of nonperfect squares can be approximated, looked up in tables, or found by using a calculator. You may want to keep these two in mind:

$$\sqrt{2} \approx 1.414 \qquad \sqrt{3} \approx 1.732$$

Simplifying square roots

Sometimes you will have to *simplify* square roots, or write them in simplest form. In fractions, $\frac{2}{4}$ can be reduced to $\frac{1}{2}$. In square roots, $\sqrt{32}$ can be simplified to $4\sqrt{2}$.

There are two main methods to *simplify a square root*.

Method 1: Factor the number under the $\sqrt{\ }$ into two factors, one of which is the largest possible perfect square. (Perfect squares are 1, 4, 9, 16, 25, 36, 49, …)

Method 2: Completely factor the number under the $\sqrt{\ }$ into prime factors and then simplify by bringing out any factors that came in pairs.

Example 8: Simplify $\sqrt{32}$.

Method 1

$$\sqrt{32} = \sqrt{16 \times 2}$$
$$= \sqrt{16} \times \sqrt{2}$$
$$= 4\sqrt{2}$$

Method 2

$$\sqrt{32} = \sqrt{2 \times 16}$$
$$= \sqrt{2 \times 2 \times 8}$$
$$= \sqrt{2 \times 2 \times 2 \times 4}$$
$$= \sqrt{2 \times 2 \times 2 \times 2 \times 2}$$
$$= \sqrt{2 \times 2} \times \sqrt{2 \times 2} \times \sqrt{2}$$
$$= 2 \times 2 \times \sqrt{2}$$
$$= 4\sqrt{2}$$

In Example 8, the largest perfect square is easy to see, and Method 1 probably is a faster method.

Example 9: Simplify $\sqrt{2016}$.

Method 1

$$\sqrt{2016} = \sqrt{144 \times 14}$$
$$= \sqrt{144} \times \sqrt{14}$$
$$= 12\sqrt{14}$$

Method 2

$$\sqrt{2016} = \sqrt{2 \times 1008}$$
$$= \sqrt{2 \times 2 \times 504}$$
$$= \sqrt{2 \times 2 \times 2 \times 252}$$
$$= \sqrt{2 \times 2 \times 2 \times 2 \times 126}$$
$$= \sqrt{2 \times 2 \times 2 \times 2 \times 2 \times 63}$$
$$= \sqrt{2 \times 2 \times 2 \times 2 \times 2 \times 3 \times 21}$$
$$= \sqrt{2 \times 2 \times 2 \times 2 \times 2 \times 3 \times 3 \times 7}$$
$$= \sqrt{2 \times 2} \times \sqrt{2 \times 2} \times \sqrt{3 \times 3} \times \sqrt{2 \times 7}$$
$$= 2 \times 2 \times 3 \times \sqrt{14}$$
$$= 12\sqrt{14}$$

In Example 9, it is not so obvious that the largest perfect square is 144, so Method 2 is probably the faster method.

Many square roots cannot be simplified because they are already in simplest form, such as $\sqrt{7}, \sqrt{10}$, and $\sqrt{15}$.

Grouping Symbols

There are basically three types of grouping symbols: parentheses, brackets, and braces.

Parentheses ( )

Parentheses are used to group numbers or variables. Everything inside parentheses must be done before any other operations.

Example 10: Simplify.

$$50(2 + 6) = 50(8) = 400$$

When a parenthesis is preceded by a minus sign, to remove the parentheses, change the sign of each term within the parentheses.

Example 11: Simplify.

$$6 - (-3 + a - 2b + c) =$$
$$6 + 3 - a + 2b - c =$$
$$9 - a + 2b - c$$

Brackets [ ] and braces { }

Brackets and *braces* also are used to group numbers or variables. Technically, they are used after parentheses. Parentheses are to be used first, then brackets, and then braces: $\{[( )]\}$. Sometimes, instead of brackets or braces, you will see the use of larger parentheses.

$$((3 + 4) \cdot 5) + 2$$

A number using all three grouping symbols would look like this:

$$2\{1 + [4(2 + 1) + 3]\}$$

Example 12: Simplify $2\{1 + [4(2 + 1) + 3]\}$. Notice that you work from the inside out.

$$2\{1 + [4(2 + 1) + 3]\} =$$
$$2\{1 + [4(3) + 3]\} =$$
$$2\{1 + [12 + 3]\} =$$
$$2\{1 + [15]\} =$$
$$2\{16\} = 32$$

Order of operations

If multiplication, division, powers, addition, parentheses, and so forth are all contained in one problem, the *order of operations* is as follows:

1. Parentheses
2. Exponents (or radicals)
3. Multiplication or division (in the order it occurs from left to right)
4. Addition or subtraction (in the order it occurs from left to right)

Many students find the made up word PEMDAS helpful as a memory tool. The "P" reminds you that "parentheses" are done first; the "E" reminds you that "exponents" are done next; the "MD" reminds you to "multiply or divide" in the order it occurs from left to right; and the "AS" reminds you to "add or subtract" in the order it occurs from left to right.

Also, some students remember the order using the following phrase:

Please	**E**xcuse	**M**y	**D**ear	**A**unt	**S**ally
Parentheses	**E**xponents	**M**ultiply or	**D**ivide	**A**dd or	**S**ubtract

Example 13: Simplify the following problems.

(a) $6 + 4 \times 3 =$
$\qquad 6 + 12 =$ (multiplication)
$\qquad\quad 18$ (then addition)

(b) $10 - 3 \times 6 + 10^2 + (6 + 1) \times 4 =$
$\qquad 10 - 3 \times 6 + 10^2 + (7) \times 4 =$ (parentheses first)
$\qquad 10 - 3 \times 6 + 100 + (7) \times 4 =$ (exponents next)
$\qquad\quad 10 - 18 + 100 + 28 =$ (multiplication)

$$-8 + 100 + 28 = \quad \text{(addition/subtraction left to right)}$$
$$92 + 28 = 120$$

Divisibility Rules

The following set of rules can help you save time in trying to check the divisibility of numbers.

A number is divisible by	if
2	it ends in 0, 2, 4, 6, or 8
3	the sum of its digits is divisible by 3
4	the number formed by the last two digits is divisible by 4
5	it ends in 0 or 5
6	it is divisible by 2 and 3 (use the rules for both)
7	(no simple rule)
8	the number formed by the last three digits is divisible by 8
9	the sum of its digits is divisible by 9

Example 14:

   (a) Is 126 divisible by 3? Sum of digits = 9. Because 9 is divisible by 3, then 126 is divisible by 3.

   (b) Is 1,648 divisible by 4? Because 48 is divisible by 4, then 1,648 is divisible by 4.

   (c) Is 186 divisible by 6? Because 186 ends in 6, it is divisible by 2. Sum of digits = 15. Because 15 is divisible by 3, 186 is divisible by 3. 186 is divisible by 2 and 3; therefore, it is divisible by 6.

   (d) Is 2,488 divisible by 8? Because 488 is divisible by 8, then 2,488 is divisible by 8.

   (e) Is 2,853 divisible by 9? Sum of digits = 18. Because 18 is divisible by 9, then 2,853 is divisible by 9.

Chapter Check-Out

Q&A

1. Which of the following are integers? 3, 4, $\frac{1}{2}$, 0, –1, 2
2. Which of the following are prime numbers? 2, 5, 7, 9, 15, 21
3. The identity element in addition is _____.
4. True or false: $a(b + c) = (a)(b)(c)$
5. $4^0 =$
6. $3^{-5} =$
7. $4^5 \times 4^8 =$ _____ (with exponents)
8. $(3^2)^3 =$ _____ (with exponents)
9. $\sqrt[3]{64} =$
10. Approximate: $\sqrt{54}$ (to the nearest tenth)
11. Simplify: $3[10(4 + 3^2)]$
12. The number 6,321 is divisible by which numbers between 1 and 10?

Answers: 1. 3, 4, 0, –1, 2 **2.** 2, 5, 7 **3.** 0 **4.** False **5.** 1 **6.** $\frac{1}{243}$ **7.** 4^{13} **8.** 3^6 **9.** 4 **10.** about 7.3 **11.** 390 **12.** 3, 7

CHAPTER 2

SIGNED NUMBERS, FRACTIONS, AND PERCENTS

Chapter Check-In

❑ Positive and negative numbers

❑ Fractions

❑ Decimals

❑ Repeating decimals

❑ Percent

❑ Scientific notation

As you get closer to entering the world of algebra, you should have a solid background in working with signed numbers, fractions, and percents.

Signed Numbers (Positive Numbers and Negative Numbers)

The term *signed numbers* refers to positive and negative numbers. If no sign is shown, the number automatically is considered positive.

Number lines

On a **number line**, numbers to the right of 0 are positive. Numbers to the left of 0 are negative, as shown in Figure 2–1.

Figure 2–1 A number line using integers.

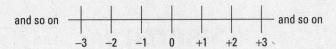

Given any two numbers on a number line, the one on the right is always larger, regardless of its sign (positive or negative). Note that fractions may also be placed on a number line as shown in Figure 2–2.

Figure 2–2 A number line using fractions.

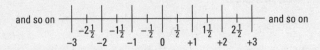

Addition of signed numbers

When *adding two numbers* with the *same sign* (either both positive or both negative), add the absolute values (the number without a sign attached) and keep the same sign. Addition problems can be presented in either a vertical form (up and down) or in a horizontal form (across).

Example 1: Add the following.

(a)
$$\begin{array}{r} +5 \\ +(+7) \\ \hline ? \end{array}$$
Add the absolute values $(5+7=12)$ and keep the sign $(+)$.

$$\begin{array}{r} +5 \\ +(+7) \\ \hline +12 \end{array}$$

(b) $-8+(-3)=?$ Add the absolute values $(8+3=11)$ and keep the sign $(-)$.

$$-8+(-3)=-11$$

When adding two numbers with different signs (one positive and one negative), subtract the absolute values and keep the sign of the one with the larger absolute value.

Example 2: Add the following.

(a)
$$+5$$
$$+(-7)$$
$$\overline{}$$

? Subtract the absolute values $(7 - 5 = 2)$ and
keep the sign of the one with the larger absolute value (-7).

$$+5$$
$$+(-7)$$
$$\overline{-2}$$

(b) $-59 + (+72) = ?$ Subtract the absolute values $(72 - 59 = 13)$ and
keep the sign of the one with the larger absolute value $(+72)$.

$-59 + (+72) = +13$

Example 3: Add the following.

(a) $\begin{array}{r} +9 \\ +(+6) \\ \hline \end{array}$ (b) $-12 + 9$ (c) $\begin{array}{r} 8 \\ +(-5) \\ \hline \end{array}$ (d) $23 + (-18)$

Answers: (a) 15 **(b)** –3 **(c)** 3 **(d)** 5

Subtraction of signed numbers

To *subtract positive and/or negative numbers,* just change the sign of the number being subtracted and then add.

Example 4: Subtract the following.

(a) $\begin{array}{r} +12 \\ -(+4) \\ \hline \end{array}$ (b) $\begin{array}{r} -14 \\ -(-4) \\ \hline \end{array}$ (c) $\begin{array}{r} -19 \\ -(+6) \\ \hline \end{array}$ (d) $\begin{array}{r} +20 \\ -(-3) \\ \hline \end{array}$

$\begin{array}{r} +12 \\ +(-4) \\ \hline +8 \end{array}$ $\begin{array}{r} -14 \\ +(+4) \\ \hline -10 \end{array}$ $\begin{array}{r} -19 \\ +(-6) \\ \hline -25 \end{array}$ $\begin{array}{r} +20 \\ +(+3) \\ \hline +23 \end{array}$

Subtracting positive and/or negative numbers may also be done "horizontally."

Example 5: Subtract the following.

  (a) $+12 - (+4) = +12 + (-4) = 8$
  (b) $+16 - (-6) = +16 + (+6) = 22$
  (c) $-20 - (+3) = -20 + (-3) = -23$
  (d) $-5 - (-2) = -5 + (+2) = -3$

Minus preceding parenthesis

If a *minus precedes a parenthesis,* it means everything within the parentheses is to be subtracted. Therefore, using the same rule as in subtraction of signed numbers, simply change every sign within the parentheses to its opposite and then add.

Example 6: Subtract the following.

  (a) $9 - (+3 - 5 + 7 - 6) =$
   $9 + (-3 + 5 - 7 + 6) =$
     $9 + (+1) = 10$

  (b) $20 - (+35 - 50 + 100) =$
   $20 + (-35 + 50 - 100) =$
     $20 + (-85) = -65$

Multiplying and dividing signed numbers

To *multiply or divide signed numbers,* treat them just like regular numbers but remember this rule: An odd number of negative signs will produce a negative answer. An even number of negative signs will produce a positive answer.

Example 7: Multiply or divide the following.

  (a) $(-3)(+8)(-5)(-1)(-2) = +240$
  (b) $(-3)(+8)(-1)(-2) = -48$
  (c) $\dfrac{-64}{-2} = +32$
  (d) $\dfrac{-64}{+2} = -32$

Fractions

A **fraction,** or fractional number, is used to represent a part of a whole. Fractions consist of two numbers: a **numerator** (which is above the line) and a **denominator** (which is below the line).

$$\frac{1}{2} \frac{\text{numerator}}{\text{denominator}}$$

The denominator tells you the number of equal parts into which something is divided. The numerator tells you how many of these equal parts are being considered. Thus, if the fraction is $\frac{3}{5}$ of a pie, the denominator 5 tells you that the pie has been divided into five equal parts, of which 3 (numerator) are in the fraction. Sometimes, it helps to think of the *dividing line* (the middle of a fraction) as meaning "out of." In other words, $\frac{3}{5}$ also means 3 out of 5 equal parts from the whole pie.

Negative fractions

Fractions may be *negative* as well as positive. (See the number line in Figure 2–2.) However, negative fractions are typically written as follows:

$$-\frac{3}{4} \text{ not } \frac{-3}{4} \text{ or } \frac{3}{-4} \text{ (although they are all equal)}$$

$$-\frac{3}{4} = \frac{-3}{4} = \frac{3}{-4}$$

Adding positive and negative fractions

The rule for adding signed numbers applies to fractions as well.

Example 8: Add the following.

(a) $\quad -\dfrac{1}{2} + \dfrac{1}{3} = -\dfrac{3}{6} + \dfrac{2}{6} = -\dfrac{1}{6}$

(b)
$$\begin{aligned}
+\frac{3}{4} &= \quad +\frac{9}{12} \\
+\left(-\frac{1}{3}\right) &= \ +\left(-\frac{4}{12}\right) \\
\hline
& \quad\quad +\frac{5}{12}
\end{aligned}$$

Subtracting positive and negative fractions

The rule for subtracting signed numbers applies to fractions as well.

Example 9: Subtract the following.

(a)
$$+\frac{9}{10} = +\frac{9}{10} = +\frac{9}{10}$$
$$-\left(-\frac{1}{5}\right) = +\frac{1}{5} = +\frac{2}{10}$$
$$+\frac{11}{10} = 1\frac{1}{10}$$

(b) $+\frac{2}{3} - \left(-\frac{1}{5}\right) = \frac{10}{15} - \left(-\frac{3}{15}\right) = \frac{10}{15} + \left(+\frac{3}{15}\right) = \frac{13}{15}$

(c) $+\frac{1}{3} - \frac{3}{4} = +\frac{4}{12} - \frac{9}{12} = +\frac{4}{12} + \left(-\frac{9}{12}\right) = -\frac{5}{12}$

Multiplying fractions

To *multiply fractions,* simply multiply the numerators and then multiply the denominators. Reduce to lowest terms if necessary.

Example 10: Multiply.

$$\frac{2}{3} \times \frac{5}{12} = \frac{10}{36} \qquad \text{reduce } \frac{10}{36} \text{ to } \frac{5}{18}$$

This answer had to be reduced because it wasn't in lowest terms. Because whole numbers can also be written as fractions $\left(3 = \frac{3}{1}, 4 = \frac{4}{1}\right)$, and so forth, the problem $3 \times \frac{3}{8}$ would be worked by changing 3 to $\frac{3}{1}$.

Early reducing

Early reducing when multiplying fractions would have eliminated the need to reduce your answers after completing the multiplication. To reduce, find a number that divides evenly into one numerator and one denominator. In this case, 2 will divide evenly into the numerator 2 (it goes in one time) and into the denominator 12 (it goes in six times). Thus,

$$\frac{\overset{1}{\cancel{2}}}{3} \times \frac{5}{\underset{6}{\cancel{12}}} = \frac{5}{18}$$

Remember, you may only do early reducing when *multiplying* fractions. The rules for multiplying signed numbers hold here, too.

Example 11: Reduce early where possible and then multiply.

(a) $\dfrac{1}{4} \times \dfrac{2}{7} = \dfrac{1}{\underset{2}{\cancel{4}}} \times \dfrac{\overset{1}{\cancel{2}}}{7} = \dfrac{1}{14}$

(b) $\left(-\dfrac{3}{8}\right) \times \left(-\dfrac{4}{9}\right) = \left(-\dfrac{\overset{1}{\cancel{3}}}{\underset{2}{\cancel{8}}}\right) \times \left(-\dfrac{\overset{1}{\cancel{4}}}{\underset{3}{\cancel{9}}}\right) = +\dfrac{1}{6}$

Multiplying mixed numbers

To *multiply mixed numbers,* first change any mixed number to an improper fraction. Then multiply as shown earlier in this chapter.

Example 12: Multiply.

$$3\dfrac{1}{3} \times 2\dfrac{1}{4} = \dfrac{10}{3} \times \dfrac{9}{4} = \dfrac{\overset{5}{\cancel{10}}}{\underset{1}{\cancel{3}}} \times \dfrac{\overset{3}{\cancel{9}}}{\underset{2}{\cancel{4}}} = \dfrac{15}{2} \ \text{ or } \ 7\dfrac{1}{2}$$

Change the answer, if in improper fraction form, back to a mixed number and reduce if necessary. Remember, the rules for multiplication of signed numbers apply here as well.

Dividing fractions

To *divide fractions,* invert (turn upside down) the second fraction (the one "divided by") and multiply. Then reduce if possible.

Example 13: Divide.

(a) $\dfrac{1}{6} \div \dfrac{1}{5} = \dfrac{1}{6} \times \dfrac{5}{1} = \dfrac{5}{6}$

(b) $\dfrac{1}{9} \div \dfrac{1}{3} = \dfrac{1}{\cancel{9}_3} \times \dfrac{\cancel{3}^1}{1} = \dfrac{1}{3}$

Here, too, the rules for division of signed numbers apply.

Dividing complex fractions

Sometimes a division of fractions problem may appear in the following form (these are called *complex fractions*).

Example 14: Simplify.

$$\dfrac{\quad\dfrac{3}{4}\quad}{\dfrac{7}{8}}$$

Consider the line separating the two fractions to mean "divided by." Therefore, this problem can be rewritten as follows:

$$\dfrac{3}{4} \div \dfrac{7}{8}$$

Now, follow the same procedure as shown in Example 13.

$$\dfrac{3}{4} \div \dfrac{7}{8} = \dfrac{3}{\cancel{4}_1} \times \dfrac{\cancel{8}^2}{7} = \dfrac{6}{7}$$

Dividing mixed numbers

To *divide mixed numbers,* first change them to improper fractions (see the section "Multiplying mixed numbers" earlier in the chapter). Then follow the rule for dividing fractions (see the "Dividing fractions" section earlier in the chapter).

Example 15: Divide.

$$3\frac{3}{5} \div 2\frac{2}{3} = \frac{18}{5} \div \frac{8}{3} = \frac{\overset{9}{\cancel{18}}}{5} \times \frac{3}{\underset{4}{\cancel{8}}} = \frac{27}{20} \text{ or } 1\frac{7}{20}$$

Notice that after you invert and have a multiplication of fractions problem, you then may do early reducing when appropriate.

Simplifying Fractions and Complex Fractions

If either numerator or denominator consists of several numbers, these numbers must be combined into one number. Then reduce if possible.

Example 16: Simplify.

(a) $\dfrac{28+14}{26+17} = \dfrac{42}{43}$

(b) $\dfrac{\dfrac{1}{4}+\dfrac{1}{2}}{\dfrac{1}{3}+\dfrac{1}{4}} = \dfrac{\dfrac{1}{4}+\dfrac{2}{4}}{\dfrac{4}{12}+\dfrac{3}{12}} = \dfrac{\dfrac{3}{4}}{\dfrac{7}{12}} = \dfrac{3}{4} \div \dfrac{7}{12} = \dfrac{3}{\underset{1}{\cancel{4}}} \times \dfrac{\overset{3}{\cancel{12}}}{7} = \dfrac{9}{7} \text{ or } 1\dfrac{2}{7}$

(c) $\dfrac{3-\dfrac{3}{4}}{-4+\dfrac{1}{2}} = \dfrac{2\dfrac{1}{4}}{-3\dfrac{1}{2}} = \dfrac{\dfrac{9}{4}}{-\dfrac{7}{2}} = \dfrac{9}{4} \div \left(-\dfrac{7}{2}\right) = \dfrac{9}{\underset{2}{\cancel{4}}} \times \left(-\dfrac{\overset{1}{\cancel{2}}}{7}\right) = -\dfrac{9}{14}$

(d) $\dfrac{1}{1+\dfrac{1}{1+\dfrac{1}{1+\dfrac{1}{4}}}} = \dfrac{1}{1+\dfrac{1}{1+\dfrac{1}{\dfrac{5}{4}}}} = \dfrac{1}{1+\dfrac{1}{1+\left(1 \div \dfrac{5}{4}\right)}} = \dfrac{1}{1+\dfrac{1}{1+\left(1 \times \dfrac{4}{5}\right)}} = \dfrac{1}{1+\dfrac{1}{1+\dfrac{4}{5}}} =$

$\dfrac{1}{1\dfrac{4}{5}} = \dfrac{1}{\dfrac{9}{5}} = 1 \div \dfrac{9}{5} = 1 \times \dfrac{5}{9} = \dfrac{5}{9}$

Decimals

Fractions also may be written in *decimal* form (decimal fractions) as either terminating, coming to an end (for example 0.3 or 0.125), or having an *infinite* (never ending) *repeating* pattern (for example (0.666... or 0.1272727...).

Changing terminating decimals to fractions

To *change terminating decimals to fractions,* remember that all numbers to the right of the decimal point are fractions with denominators of only 10, 100, 1000, 10,000, and so forth. Next, use the technique of *read it, write it,* and *reduce it.*

Example 17: Change the following to fractions in lowest terms.

(a) 0.8

Read it: eight tenths

Write it: $\dfrac{8}{10}$

Reduce it: $\dfrac{4}{5}$

(b) -0.07

Read it: negative seven hundredths

Write it: $-\dfrac{7}{100}$ (can't reduce this one)

All rules for signed numbers also apply to operations with decimals.

Changing fractions to decimals

To *change a fraction to a decimal,* simply do what the operation says. In other words $\dfrac{13}{20}$ means 13 divided by 20. So do just that (insert decimal points and zeros accordingly).

Example 18: Change to decimals.

(a) $\dfrac{13}{20}$

$$20\overline{)13.00} = 0.65$$

with quotient 0.65

(b) $-\dfrac{2}{9}$

$$9\overline{)2.00000} = -0.222\ldots$$

with quotient $-0.222\ldots$

Changing infinite repeating decimals to fractions

Infinite repeating decimals usually are represented by putting a line over (sometimes under) the shortest block of repeating decimals. This line is called a *vinculum*. So you would write

$$0.\overline{3} \text{ to indicate } .333\ldots$$

$$0.\overline{51} \text{ to indicate } .515151\ldots$$

$$-2.1\overline{47} \text{ to indicate } -2.1474747\ldots$$

Notice that only the digits under the vinculum are repeated.

Every infinite repeating decimal can be expressed as a fraction.

Example 19: Find the fraction represented by the repeating decimal $0.\overline{7}$.

Let n stand for $\quad 0.\overline{7}$ or $0.7777\ldots$

So $10n$ stands for $7.\overline{7}$ or $7.7777\ldots$

Because $10n$ and n have the same fractional part, their difference is an integer.

$$10n = 7.\overline{7}$$
$$-n = -0.\overline{7}$$
$$\overline{}$$
$$9n = 7$$

$$n = \frac{7}{9}$$

Therefore, $0.\overline{7} = \dfrac{7}{9}$

Example 20: Find the fraction represented by the repeating decimal $0.\overline{36}$.

Let n stand for $\quad 0.\overline{36}$ or $0.363636...$

So $100n$ stands for $36.\overline{36}$ or $36.363636...$

Because $100n$ and n have the same fractional part, their difference is an integer.

$$100n = 36.\overline{36}$$
$$-n = -0.\overline{36}$$
$$\overline{ }$$
$$99n = 36$$

$$n = \frac{36}{99} = \frac{4}{11}$$

Therefore, $\quad 0.\overline{36} = \frac{4}{11}$

Example 21: Find the fraction represented by the repeating decimal $0.5\overline{4}$.

Let n stand for $\quad 0.5\overline{4}$ or $0.54444...$

So $10n$ stands for $\quad 5.\overline{4}$ or $5.4444...$

And $100n$ stands for $54.\overline{4}$ or $54.4444...$

Because $100n$ and $10n$ have the same fractional part, their difference is an integer.

$$100n = 54.\overline{4}$$
$$-10n = -5.\overline{4}$$
$$\overline{ }$$
$$90n = 49$$

$$n = \frac{49}{90}$$

Therefore, $\quad 0.5\overline{4} = \frac{49}{90}$

Percent

A fraction whose denominator is 100 is called a *percent*. The word *percent* means hundredths (per hundred).

So $\qquad\qquad 37\% = \dfrac{37}{100}$

Changing decimals to percents

To change decimals to percents,

1. **Move the decimal point two places to the right.**
2. **Insert a percent sign.**

Example 22: Change to percents.

    (a) 0.09 = 9%
    (b) 0.75 = 75%
    (c) 1.85 = 185%
    (d) 0.002 = 0.2%
    (e) 8.7 = 870%

Changing percents to decimals

To change percents to decimals,

1. **Eliminate the percent sign.**
2. **Move the decimal point two places to the left (sometimes, adding zeros will be necessary).**

Example 23: Change to decimals.

    (a) 23% = 0.23
    (b) 5% = 0.05
    (c) 0.7% = 0.007
    (d) $16\frac{2}{3}\% = 0.16\frac{2}{3}$

Changing fractions to percents

There are two simple methods for changing fractions to percents.

Method 1

1. **Change to a decimal.**
2. **Change the decimal to a percent.**

Example 24a (Applying method 1): Change to percents.

(a) $\dfrac{2}{5} = 0.4 = 40\%$

(b) $\dfrac{5}{2} = 2.5 = 250\%$

(c) $\dfrac{1}{20} = 0.05 = 5\%$

Method 2

1. **Make a proportion with the fraction on the left side of the equal sign and $\dfrac{x}{100}$ on the right side of the equal sign.**

2. **Solve the proportion for x, then place a % sign to its right.**

Example 24b (Applying method 2): Change to percents.

(a) $\dfrac{3}{4} \rightarrow \dfrac{3}{4} = \dfrac{x}{100}$

$\quad 4x = 300$

$\quad x = \dfrac{300}{4} = 75$ Therefore, $\dfrac{3}{4} = 75\%$.

(b) $\dfrac{2}{3} \rightarrow \dfrac{2}{3} = \dfrac{x}{100}$

$\quad 3x = 200$

$\quad x = \dfrac{200}{3} = 66\dfrac{2}{3}$ Therefore, $\dfrac{2}{3} = 66\dfrac{2}{3}\%$.

(c) $\dfrac{1}{20} \rightarrow \dfrac{1}{20} = \dfrac{x}{100}$

$\quad 20x = 100$

$\quad x = \dfrac{100}{20} = 5$ Therefore, $\dfrac{1}{20} = 5\%$.

Changing percents to fractions

There are two simple methods for changing percents to fractions.

Method 1
1. **Drop the percent sign.**
2. **Write over one hundred.**
3. **Reduce if necessary.**

Example 25a (Applying method 1): Change to fractions.

(a) $60\% = \dfrac{60}{100} = \dfrac{3}{5}$

(b) $230\% = \dfrac{230}{100} = \dfrac{23}{10}\left(\text{or } 2\dfrac{3}{10}\right)$

Method 2
1. **Drop the percent sign.**
2. **Multiply by $\dfrac{1}{100}$.**
3. **Reduce if necessary.**

Example 25b (Applying method 2): Change to fractions.

(a) $66\dfrac{2}{3}\% \rightarrow 66\dfrac{2}{3} \times \dfrac{1}{100} = \dfrac{\overset{2}{\cancel{200}}}{3} \times \dfrac{1}{\underset{1}{\cancel{100}}} = \dfrac{2}{3}$

(b) $112\dfrac{1}{2}\% \rightarrow 112\dfrac{1}{2} \times \dfrac{1}{100} = \dfrac{\overset{9}{\cancel{225}}}{2} \times \dfrac{1}{\underset{4}{\cancel{100}}} = \dfrac{9}{8} \text{ or } 1\dfrac{1}{8}$

Memorizing the following can eliminate computations:

$$\frac{1}{100} = 0.01 = 1\% \qquad \frac{1}{8} = 0.125 = 12.5\% = 12\frac{1}{2}\% \qquad \frac{1}{6} = 0.16\frac{2}{3} = 16\frac{2}{3}\%$$

$$\frac{1}{10} = 0.1 = 10\% \qquad \frac{2}{8} = \frac{1}{4} = 0.25 = 25\% \qquad \frac{2}{6} = \frac{1}{3} = 0.33\frac{1}{3} = 33\frac{1}{3}\%$$

$$\frac{2}{10} = \frac{1}{5} = 0.2 = 20\% \qquad \frac{3}{8} = 0.375 = 37.5\% = 37\frac{1}{2}\% \qquad \frac{3}{6} = \frac{1}{2} = 0.5 = 50\%$$

$$\frac{3}{10} = 0.3 = 30\% \qquad \frac{4}{8} = \frac{1}{2} = 0.5 = 50\% \qquad \frac{4}{6} = \frac{2}{3} = 0.66\frac{2}{3} = 66\frac{2}{3}\%$$

$$\frac{4}{10} = \frac{2}{5} = 0.4 = 40\% \qquad \frac{5}{8} = 0.625 = 62.5\% = 62\frac{1}{2}\% \qquad \frac{5}{6} = 0.83\frac{1}{3} = 83\frac{1}{3}\%$$

$$\frac{5}{10} = \frac{1}{2} = 0.5 = 50\% \qquad \frac{6}{8} = \frac{3}{4} = 0.75 = 75\%$$

$$\frac{6}{10} = \frac{3}{5} = 0.6 = 60\% \qquad \frac{7}{8} = 0.875 = 87.5\% = 87\frac{1}{2}\%$$

$$\frac{7}{10} = 0.7 = 70\%$$

$$\frac{8}{10} = \frac{4}{5} = 0.8 = 80\%$$

$$\frac{9}{10} = 0.9 = 90\%$$

$$1 = 1.0 = 100\%$$

$$2 = 2.0 = 200\%$$

$$3\frac{1}{2} = 3.5 = 350\%$$

Finding the percent of a number

To *determine the percent of a number,* change the percent to a fraction or decimal (whichever is easier for you) and multiply. Remember, the word *of* means multiply.

Example 26: Find the percents of these numbers.

(a) 20% of 80 =

$$\frac{\overset{1}{\cancel{20}}}{\underset{5}{\cancel{100}}} \times \frac{80}{1} = \frac{80}{5} = 16 \quad \text{or} \quad 0.20 \times 80 = 16.00 = 16$$

(b) $\frac{1}{2}$% of 18 =

$$\frac{\frac{1}{2}}{100} \times \frac{18}{1} = \frac{1}{\underset{100}{\cancel{200}}} \times \frac{\overset{9}{\cancel{18}}}{1} = \frac{9}{100} \quad \text{or} \quad 0.005 \times 18 = 0.09$$

Other applications of percent

Turn the question word-for-word into an equation. For *what*, substitute the letter *x*; for *is*, substitute an *equal sign;* for *of*, substitute a *multiplication sign*. Change percents to decimals or fractions, whichever you find easier. Then solve the equation.

Example 27: Turn each of the following into an equation and solve.

(a) 18 is what percent of 90?

$$18 = (x)(90)$$

$$\frac{18}{90} = x$$

$$\frac{1}{5} = x$$

$$20\% = x \qquad \text{Therefore, 18 is 20\% of 90.}$$

(b) 10 is 50 percent of what number?

$$10 = (0.50)(x)$$

$$\frac{10}{0.50} = x$$

$$20 = x \qquad \text{Therefore, 10 is 50\% of 20.}$$

(c) What is 15 percent of 60?

$$x = \frac{\overset{3}{\cancel{15}}}{\underset{\underset{1}{20}}{\cancel{100}}} \times \frac{\overset{3}{\cancel{60}}}{1} = 9 \qquad \text{or} \qquad x = 0.15(60) = 9$$

Therefore, 9 is 15% of 60.

Percent–proportion method

Another simple method commonly used to solve percent problems is the **proportion** or *is/of method*. First set up a blank proportion and then fill in the empty spaces by using the following steps.

This method, using proportions with is/of, works for the three basic types of percent questions.

$$\frac{\text{\%-number}}{100} = \frac{\text{"is"-number}}{\text{"of"-number}}$$

Example 28a: 30 is what percent of 50?

The "is" number is 30. The % number is unknown, so call it x.

The "of" number is 50. Now replace these values in the appropriate positions in the proportion and solve for x.

$$\frac{x}{100} = \frac{30}{50}$$

In this particular instance, it can be observed that $\frac{60}{100} = \frac{30}{50}$ so the answer is 60%. Solving mechanically on this problem would not be time effective.

Example 28b: 30 is 20% of what number?

$$\frac{20}{100} = \frac{30}{x}$$
$$20x = 3000$$
$$x = 150$$

Therefore, 30 is 20% of 150.

Example 28c: What number is 30% of 50?

$$\frac{30}{100} = \frac{x}{50}$$
$$100x = 1500$$
$$x = 15$$

Therefore, 15 is 30% of 50.

Scientific Notation

Very large or very small numbers are sometimes written in *scientific notation*. A number written in scientific notation is a decimal number between 1 and 10 multiplied by a power of 10.

Example 29: Express the following in scientific notation.

(a) 2,100,000 written in scientific notation is 2.1×10^6. Simply place the decimal point to get a number between 1 and 10 and then count the digits to the right of the decimal to get the power of 10.

2.100000.   moved 6 digits to the left

(b) 0.0000004 written in scientific notation is 4.0×10^{-7}. Simply place the decimal point to get a number between 1 and 10 and then count the digits from the original decimal point to the new one.

.0000004.   moved 7 digits to the right

Notice that number values greater than 1 have positive exponents as the power of 10 and that number values between 0 and 1 have negative exponents as the power of 10.

Multiplication in scientific notation

To *multiply* numbers in *scientific notation*, multiply the numbers that are between 1 and 10 together to get a whole number. Then add the exponents on the 10's to get a new exponent on 10. It may be necessary to make adjustments to this answer in order to correctly express it in scientific notation.

Example 30: Multiply and express the answers in scientific notation.

(a) $(2 \times 10^2)(3 \times 10^4) = (2 \times 3) \times 10^{2+4} = 6 \times 10^6$

(b) $(6 \times 10^5)(5 \times 10^7) = (6 \times 5) \times 10^{5+7} = 30 \times 10^{12}$

This answer must be changed to scientific notation (first number between 1 and 10):

$$30 \times 10^{12} = 3.0 \times 10^1 \times 10^{12} = 3.0 \times 10^{13}$$

(c) $(4 \times 10^{-4})(2 \times 10^5) = (4 \times 2) \times 10^{-4+5} = 8 \times 10^1$

Division in scientific notation

To *divide* numbers in *scientific notation,* divide the numbers that are between 1 and 10 to get a decimal number. Then subtract the exponents on the 10s to get a new exponent on 10. It may be necessary to make adjustments to this answer in order to correctly express it in scientific notation.

Example 31: Divide and express the answers in scientific notation.

(a) $\left(8\times10^5\right)\div\left(2\times10^2\right)=(8\div2)\times10^{5-2}=4\times10^3$

(b) $\dfrac{7\times10^9}{4\times10^3}=(7\div4)\times10^{9-3}=1.75\times10^6$

(c) $\left(6\times10^7\right)\div\left(3\times10^9\right)=(6\div3)\times10^{7-9}=2\times10^{-2}$

(d) $\left(2\times10^4\right)\div\left(5\times10^2\right)=(2\div5)\times10^{4-2}=0.4\times10^2$

This answer must be changed to scientific notation.

$$0.4\times10^2=4\times10^{-1}\times10^2=4\times10^1$$

(e) $\left(8.4\times10^5\right)\div\left(2.1\times10^{-4}\right)=(8.4\div2.1)\times10^{5-(-4)}=4\times10^9$

Chapter Check-Out

1. $-12 + 9 =$

2. $(-2)(-3)(6) =$

3. $-\dfrac{1}{4} + \dfrac{2}{8}$

4. $\dfrac{3}{5} \times \dfrac{25}{36}$

5. $5\dfrac{1}{5} \div 2\dfrac{1}{6}$

6. Change to decimal: $\dfrac{1}{8}$

7. Change to fraction: $0.\overline{8}$

8. 15 is what percent of 60?

9. $(3 \times 10^4)(2 \times 10^8) =$

Answers: 1. -3 **2.** 36 **3.** 0 **4.** $\dfrac{5}{12}$ **5.** $\dfrac{12}{5}$ or $2\dfrac{2}{5}$ **6.** 0.125 **7.** $\dfrac{8}{9}$ **8.** 25%
9. 6×10^{12}

CHAPTER 3

TERMINOLOGY, SETS, AND EXPRESSIONS

Chapter Check-In

❑ Set theory

❑ Algebraic expressions

❑ Evaluating expressions

Understanding the language of algebra and how to work with algebraic expressions gives you a good foundation for learning the rules of algebra.

Set Theory

A **set** is a group of objects, numbers, and so forth. {1,2,3} is a set consisting of the numbers 1,2, and 3. Verbally, "3 is an element of the set {1,2,3}." To show this symbolically, use the symbol ∈, which is read as "is an element of" or "is a member of." Therefore, you could have written:

$$3 \in \{1,2,3\}$$

Special sets

A **subset** is a set contained within another set, or it can be the entire set itself. The set {1,2} is a subset of the set {1,2,3}, and the set {1,2,3} is a subset of the set {1,2,3}. When the subset is missing some elements that are in the set it is being compared to, it is a **proper subset.** When the subset is the set itself, it is an **improper subset.** The symbol used to indicate "is a proper subset of" is ⊂. When there is the possibility of using an improper subset, the symbol used is ⊆. Therefore, {1,2} ⊂ {1,2,3} and

{1,2,3} ⊆ {1,2,3}. The **universal set** is the general category set, or the set of all those elements under consideration. The **empty set,** or **null set,** is the set with no elements or *members*. The empty set, or null set, is represented by ∅, or { }. However, it is never represented by {∅}.

Both the universal set and the empty set are subsets of every set.

Describing sets

Rule is a method of naming a set by describing its elements.

{$x:x > 3$, x is a whole number} describes the set with elements 4, 5, 6,…. Therefore, {$x:x > 3$, x is a whole number} is the same as {4,5,6,…}. {$x:x > 3$} describes all numbers greater than 3. This set of numbers cannot be represented as a list and is represented using a number line graph, which is discussed in Chapter 8.

Roster is a method of naming a set by listing its members.

{1,2,3} is the set consisting of only the elements 1,2, and 3. There are many ways to represent this set using a rule. Two correct methods are as follows:

$$\{x:x < 4, x \text{ is a natural number}\}$$

$$\{x:0 < x < 4, x \text{ is a whole number}\}$$

An incorrect method would be {$x:0 < x < 4$} because this rule includes ALL numbers between 0 and 4, not just the numbers 1, 2, and 3.

Types of sets

Finite sets have a countable number of elements. For example, {a,b,c,d,e} is a set of five elements, thus it is a finite set. **Infinite sets** contain an uncountable number of elements. For example, {1,2,3, …} is a set with an infinite number of elements, thus it is an infinite set.

Comparing sets

Equal sets are those that have the exact same members — {1, 2, 3} = {3, 2, 1}. **Equivalent sets** are sets that have the same number of members — {1, 2, 3} ~ {a, b, c}.

Venn diagrams (and *Euler circles*) are ways of pictorially describing sets as shown in Figure 3–1.

Figure 3–1 A Venn diagram

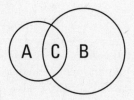

The *A* represents all the elements in the smaller oval; the *B* represents all the elements in the larger oval; and the *C* represents all the elements that are in both ovals at the same time.

Operations with sets

The **union** of two sets is a set containing all the numbers in those sets, but any duplicates are only written once. The symbol for finding the union of two sets is ∪.

Example 1. Find the union {1,2,3} ∪ {3,4,5}.

$$\{1,2,3\} \cup \{3,4,5\} = \{1,2,3,4,5\}$$

The union of the set with members 1, 2, 3 together with the set with members 3, 4, 5 is the set with members 1, 2, 3, 4, 5.

The **intersection** of two sets is a set containing only the members that are in each set at the same time. The symbol for finding the intersection of two sets is ∩.

Example 2. Find the intersection {1,2,3} ∩ {3,4,5}.

$$\{1,2,3\} \cap \{3,4,5\} = \{3\}$$

The intersection of the set with members 1, 2, 3 together with the set with members 3, 4, 5 is the set that has only the 3.

If you were to let the set with {1,2,3} be set *A*, and the set with {3,4,5} be set *B*, then you could use Venn diagrams to illustrate the situation (see Figure 3–2).

Figure 3–2 Intersection of set A and set B

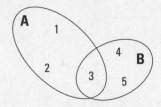

The union will be all the numbers represented in the diagram, {1,2,3,4,5}. The intersection would be where the two ovals overlap in the diagram, {3}.

Example 3. Find {1,2,3} ∩ {4,5}.

Since there are no members that are in both sets at the same time, then {1,2,3} ∩ {4,5} = ∅.

The intersection of the set with members 1, 2, 3 together with the set with members 4, 5 is the empty set, or null set. There are no members in both sets at the same time.

Variables and Algebraic Expressions

A **variable** is a symbol used to denote any element of a given set—often a letter used to stand for a number. Variables are used to change verbal expressions into **algebraic expressions.**

Example 4: Give the algebraic expression.

Verbal Expression	Algebraic Expression
The sum of a number n and 7	$n + 7$ or $7 + n$
The number n diminished by 10	$n - 10$
Seven times the number n	$7n$
x divided by 4	$\dfrac{x}{4}$
Five more than the product of 2 and n	$2n + 5$ or $5 + 2n$
Seven less than the quotient of y and 4	$\dfrac{y}{4} - 7$

■ **Key words denoting addition:**

sum	larger than	enlarge
plus	gain	rise
more than	increase	grow
greater than		

■ **Key words denoting subtraction:**

difference	smaller than	lower
minus	fewer than	diminish
lose	decrease	reduced
less than	drop	

■ **Key words denoting multiplication:**

product	times	of
multiplied by	twice	

■ **Key words denoting division:**

quotient	ratio
divided by	half

Evaluating Expressions

To **evaluate** an *expression*, just replace the unknowns with grouping symbols, insert the **value** for the unknowns, and do the arithmetic, making sure to follow the rules for the order of operations.

Example 5: Evaluate each of the following:

(a) $ab + c$ if $a = 5$, $b = 4$, and $c = 3$

$$5(4) + 3 =$$
$$20 + 3 = 23$$

(b) $2x^2 + 3y + 6$ if $x = 2$ and $y = 9$

$$2(2)^2 + 3(9) + 6 =$$
$$2(4) + 27 + 6 =$$
$$8 + 27 + 6 =$$
$$35 + 6 = 41$$

(c) $-4p^2 + 5q - 7$ if $p = -3$ and $q = -8$

$$-4(-3)^2 + 5(-8) - 7 =$$
$$-4(9) + 5(-8) - 7 =$$
$$-36 - 40 - 7 = -83$$

(d) $\dfrac{a+c}{b} + \dfrac{a}{b+c}$ if $a = 3$, $b = -2$, and $c = 7$

$$\frac{(3)+(7)}{-2} + \frac{(3)}{(-2)+(7)} =$$

$$\frac{10}{-2} + \frac{3}{5} = -5 + \frac{3}{5} = -4\frac{2}{5}$$

(e) $5x^3y^2$ if $x = -2$ and $y = 3$

$$5(-2)^3(3)^2 =$$
$$5(-8)(9) =$$
$$-40(9) = -360$$

Chapter Check-Out

1. $\{2,4,6\} \cap \{2,3,4\} =$

2. $\{3,4,5\} \cup \{5,6,7\} =$

3. True or False: $\{5,6\} \subset \{1,3,5\}$

4. Express algebraically: six more than three times a number n.

5. Evaluate: $6x^2y$ if $x = -2$ and $y = 4$.

6. Evaluate: $\dfrac{a-b}{c} - \dfrac{a+b}{2}$ if $a = 5$, $b = -1$, and $c = 3$.

Answers: 1. $\{2,4\}$ **2.** $\{3,4,5,6,7\}$ **3.** False **4.** $3n + 6$ **5.** 96 **6.** 0

CHAPTER 4

EQUATIONS, RATIOS, AND PROPORTIONS

Chapter Check-In

❑ Axioms of equality

❑ Solving equations

❑ Solving proportions for value

Working with variables and solving equations often are considered the basis of algebra.

Equations

An **equation** is a mathematical sentence, a relationship between numbers and/or symbols that says two expressions are of equal value.

Axioms of equality

For all real numbers a, b, and c, the following are some basic rules for using the equal sign.

- **Reflexive axiom:** $a = a$.

  Therefore, $4 = 4$.

- **Symmetric axiom:** If $a = b$, then $b = a$.

  Therefore, if $2 + 3 = 5$, then $5 = 2 + 3$.

- **Transitive axiom:** If $a = b$ and $b = c$, then $a = c$.

  Therefore, if $1 + 3 = 4$ and $4 = 2 + 2$, then $1 + 3 = 2 + 2$.

- **Additive axiom of equality:** If $a = b$ and $c = d$, then $a + c = b + d$.

  Therefore, if $1 + 1 = 2$ and $3 + 3 = 6$, then $1 + 1 + 3 + 3 = 2 + 6$.

- **Multiplicative axiom of equality:** If $a = b$ and $c = d$, then $ac = bd$.

  Therefore, if $1 = \dfrac{2}{2}$ and $4 = \dfrac{8}{2}$, then $1(4) = \left(\dfrac{2}{2}\right)\left(\dfrac{8}{2}\right)$.

Solving equations

Remember that an equation is like a balance scale with the equal sign (=) being the fulcrum, or center. Thus, if you do the *same thing to each side of the equation* (say, add 5 to each side), the equation will still be balanced.

Example 1: Solve for x.

$$x - 5 = 23$$

To solve the equation $x - 5 = 23$, you must get x by itself on one side; therefore, add 5 to *each side of the equation.*

$$
\begin{array}{r}
x - 5 = 23 \\
\underline{+5 \quad +5} \\
x \quad\ = 28
\end{array}
$$

In the same manner, you may subtract, multiply, or divide both sides of an equation by the same (nonzero) number, and the solution will not change. Sometimes you may have to use more than one step to solve for an unknown.

Example 2: Solve for x.

$$3x + 4 = 19$$

Subtract 4 from each side of the equation to get the $3x$ by itself on one side.

$$
\begin{array}{r}
3x + 4 = 19 \\
\underline{-4 \quad -4} \\
3x \quad\ = 15
\end{array}
$$

Then divide each side of the equation by 3 to get x.

$$\frac{3x}{3} = \frac{15}{3}$$

$$x = 5$$

Remember that solving an equation is using opposite operations until the letter is on a side by itself (for addition, subtract; for multiplication, divide; and so forth).

To check, substitute your answer into the original equation.

$$3x + 4 = 19$$
$$3(5) + 4 = 19$$
$$15 + 4 = 19$$
$$19 = 19$$

Example 3: Solve for x.

$$\frac{x}{5} - 4 = 2$$

Add 4 to each side of the equation.

$$\frac{x}{5} - 4 = 2$$
$$\underline{+4 +4}$$
$$\frac{x}{5} = 6$$

Multiply each side of the equation by 5 to get x.

$$5\left(\frac{x}{5}\right) = 5(6)$$
$$x = 30$$

Example 4: Solve for x.

$$\frac{3}{5}x - 6 = 12$$

Add 6 to each side of the equation.

$$\frac{3}{5}x - 6 = 12$$
$$\underline{+6 +6}$$
$$\frac{3}{5}x = 18$$

Multiply each side of the equation by $\frac{5}{3}$ (same as dividing by $\frac{3}{5}$).

$$\frac{5}{3}\left(\frac{3}{5}x\right) = \frac{5}{3}\left(\frac{18}{1}\right)$$

$$x = \frac{5}{\cancel{3}}\left(\frac{\overset{6}{\cancel{18}}}{1}\right)$$

$$x = 30$$

Example 5: Solve for x.

$$5x = 2x - 6$$

Subtract $2x$ from each side of the equation.

$$
\begin{array}{rcl}
5x & = & 2x - 6 \\
-2x & & -2x \\
\hline
3x & = & -6
\end{array}
$$

Divide each side of the equation by 3.

$$\frac{3x}{3} = \frac{-6}{3}$$

$$x = -2$$

Example 6: Solve for x.

$$6x + 3 = 4x + 5$$

Subtract $4x$ from each side of the equation.

$$
\begin{array}{rcl}
6x + 3 & = & 4x + 5 \\
-4x & & -4x \\
\hline
2x + 3 & = & 5
\end{array}
$$

Subtract 3 from each side of the equation.

$$
\begin{array}{rcl}
2x + 3 & = & 5 \\
-3 & & -3 \\
\hline
2x & = & 2
\end{array}
$$

Divide each side of the equation by 2.

$$\frac{2x}{2} = \frac{2}{2}$$
$$x = 1$$

Literal equations

Literal equations have no numbers, only symbols (letters).

Example 7: Solve for q.

$$qp - x = y$$

First add x to each side of the equation.

$$
\begin{array}{rl}
qp - x &= y \\
+x & +x \\
\hline
qp &= y + x
\end{array}
$$

Then divide each side of the equation by p.

$$\frac{qp}{p} = \frac{y + x}{p}$$
$$q = \frac{y + x}{p}$$

Operations opposite to those in the original equation were used to isolate q. (To remove the $-x$, an x was added to each side of the equation. Because the problem has q times p, each side of the equation was divided by p.)

Example 8: Solve for y.

$$\frac{y}{x} = c$$

Multiply each side of the equation by x to get y alone.

$$x\left(\frac{y}{x}\right) = x(c)$$
$$y = xc$$

Example 9. Solve for x.

$$ax + b = cx + d$$

Subtract cx from each side of the equation.

$$
\begin{array}{rcl}
ax + b & = & cx + d \\
-cx & & -cx \\
\hline
ax - cx + b & = & d \\
(a-c)x + b & = & d
\end{array}
$$

$ax - cx = (a - c)x$ by the distributive property.

Subtract b from each side of the equation.

$$
\begin{array}{rcl}
(a-c)x + b & = & d \\
-b & & -b \\
\hline
(a-c)x & = & d - b
\end{array}
$$

Divide each side of the equation by $(a - c)$.

$$(a-c)x = d - b$$

$$\frac{(a-c)x}{(a-c)} = \frac{d-b}{a-c}$$

$$x = \frac{d-b}{a-c}$$

Ratios and Proportions

Ratios and proportions are not only used in arithmetic, but are also commonly used in algebra (and geometry). The definitions given in this chapter are the same as those used in arithmetic.

Ratios

A **ratio** is a method of comparing two or more numbers or variables. Ratios are written as $a{:}b$ or in working form, as a fraction.

$$\frac{a}{b} \text{ or } \frac{a}{b} \text{ is read "}a \text{ is to } b\text{"}$$

Notice that whatever comes after the "to" goes second or at the bottom of the fraction.

Proportions

Proportions are written as two ratios (fractions) equal to each other.

Example 10: Solve this problem for x.

$$p \text{ is to } q \text{ as } x \text{ is to } y$$

First the proportion may be rewritten.

$$\frac{p}{q} = \frac{x}{y}$$

Now simply multiply each side by y.

$$y\left(\frac{p}{q}\right) = y\left(\frac{x}{y}\right)$$

$$\frac{yp}{q} = x$$

Example 11: Solve this proportion for t.

$$s \text{ is to } t \text{ as } r \text{ is to } q$$

Rewrite.

$$\frac{s}{t} = \frac{r}{q}$$

Multiply each side of the equation by the common denominator, tq, and do early reducing.

$$(\cancel{t})(q)\frac{s}{\cancel{t}} = (t)(\cancel{q})\frac{r}{\cancel{q}}$$

$$qs = tr$$

Note that the result is the same as if you had "cross-multiplied" through the equal sign. In future examples, this method is referred to as "cross-multiply."

Divide each side of the equation by r.

$$\frac{qs}{r} = \frac{tr}{r}$$

$$\frac{qs}{r} = t \text{ or } t = \frac{qs}{r}$$

Solving proportions for value

Follow the procedures given in Examples 10 and 11 to solve for the unknown.

Example 12: Solve for x.

$$\frac{4}{x} = \frac{2}{5}$$

Cross multiply.

$$(4)(5) = 2x$$

$$20 = 2x$$

Divide each side of the equation by 2.

$$\frac{20}{2} = \frac{2x}{2}$$

$$10 = x$$

Example 13: Solve for x.

$$\frac{2x-4}{3x+6} = \frac{2}{5}$$

Cross multiply.

$$5(2x - 4) = 2(3x + 6)$$

Use the distributive property.

$$10x - 20 = 6x + 12$$

Subtract $6x$ from each side of the equation.

$$\begin{array}{r} 10x - 20 = 6x + 12 \\ -6x \qquad -6x \\ \hline 4x - 20 = \quad 12 \end{array}$$

Add 20 to each side of the equation.

$$\begin{array}{r} 4x - 20 = \quad 12 \\ +20 = +20 \\ \hline 4x \qquad = \quad 32 \end{array}$$

Divide each side of the equation by 4.

$$\frac{4x}{4} = \frac{32}{4}$$
$$x = 8$$

Chapter Check-Out

1. True or false: If $a = b$ and $b = c$, then $a = c$.
2. Solve for x: $\dfrac{x}{4} - 5 = 8$.
3. Solve for x: $7x + 3 = 5x + 7$
4. Solve for m: $mn - r = q$.
5. Solve for x: $\dfrac{a}{x} = \dfrac{b}{c}$.
6. Solve for y: m is to n as y is to z.
7. Solve for x: $\dfrac{6}{x} = \dfrac{3}{5}$.

Answers: 1. True **2.** 52 **3.** 2 **4.** $\dfrac{q+r}{n}$ **5.** $\dfrac{ac}{b}$ **6.** $\dfrac{mz}{n}$ **7.** 10

CHAPTER 5

EQUATIONS WITH TWO VARIABLES

Chapter Check-In

❏ Solving systems of equations

❏ Addition/subtraction method

❏ Substitution method

❏ Graphing method

If you have two different equations with the same two unknowns in each, you can solve for both unknowns.

Solving Systems of Equations (Simultaneous Equations)

There are three common methods for solving: addition/subtraction, substitution, and graphing.

Addition/subtraction method

This method is also known as the elimination method.

To use the addition/subtraction method, do the following:

1. Multiply one or both equations by some number(s) to make the number in front of one of the letters (unknowns) the same or exactly the opposite in each equation.
2. Add or subtract the two equations to eliminate one letter.
3. Solve for the remaining unknown.
4. Solve for the other unknown by inserting the value of the unknown found in one of the original equations.

Example 1: Solve for x and y.

$$x + y = 7$$
$$x - y = 3$$

Adding the equations eliminates the y-terms.

$$x + y = 7$$
$$\underline{x - y = 3}$$
$$2x \quad = 10$$
$$\frac{2x}{2} = \frac{10}{2}$$
$$x = 5$$

Now inserting 5 for x in the first equation gives the following:

$$5 + y = 7$$
$$\underline{-5 \quad\quad -5}$$
$$y = 2$$

Answer: $x = 5$, $y = 2$

By replacing each x with a 5 and each y with a 2 in the original equations, you can see that each equation will be made true.

In Example 1 and Example 2, a unique answer existed for x and y that made each sentence true at the same time. In some situations you do not get unique answers or you get no answers. You need to be aware of these when you use the addition/subtraction method.

Example 2: Solve for x and y.

$$3x + 3y = 24$$
$$2x + y = 13$$

First multiply the bottom equation by 3. Now the y is preceded by a 3 in each equation.

$$3x + 3y = 24 \qquad\qquad 3x + 3y = 24$$
$$3(2x) + 3(y) = 3(13) \qquad\qquad 6x + 3y = 39$$

The equations can be subtracted, eliminating the y terms.

$$3x + 3y = 24$$
$$\underline{-6x + (-3y) = -39}$$
$$-3x = -15$$
$$\frac{-3x}{-3} = \frac{-15}{-3}$$
$$x = 5$$

Insert $x = 5$ in one of the original equations to solve for y.

$$2x + y = 13$$
$$2(5) + y = 13$$
$$10 + y = 13$$
$$\underline{-10 -10}$$
$$y = 3$$

Answer: $x = 5$, $y = 3$

Of course, if the number in front of a letter is already the same in each equation, you do not have to change either equation. Simply add or subtract.

To check the solution, replace each x in each equation with 5 and replace each y in each equation with 3.

$$3x + 3y = 24 \qquad\qquad 2x + y = 13$$
$$3(5) + 3(3) = 24 \qquad\qquad 2(5) + (3) = 13$$
$$15 + 9 = 24 \qquad\qquad 10 + 3 = 13$$
$$24 = 24\checkmark \qquad\qquad 13 = 13\checkmark$$

Example 3. Solve for a and b.

$$3a + 4b = 2$$
$$6a + 8b = 4$$

Multiply the top equation by 2. Notice what happens.

$$2(3a) + 2(4b) = 2(2) \qquad 6a + 8b = 4$$
$$6a + 8b = 4 \qquad\qquad 6a + 8b = 4$$

Now if you were to subtract one equation from the other, the result is $0 = 0$.

This statement is *always true*.

When this occurs, the system of equations does not have a unique solution. In fact, any a and b replacement that makes one of the equations true, also makes the other equation true. For example, if $a = -6$ and $b = 5$, then both equations are made true.

$$[3(-6) + 4(5) = 2 \text{ AND } 6(-6) + 8(5) = 4]$$

What we have here is really only one equation written in two different ways. In this case, the second equation is actually the first equation multiplied by 2. The solution for this situation is either of the original equations or a simplified form of either equation.

Example 4. Solve for x and y.

$$3x + 4y = 5$$
$$6x + 8y = 9$$

Multiply the top equation by 2. Notice what happens.

$$2(3x) + 2(4y) = 2(5) \qquad 6x + 8y = 10$$
$$6x + 8y = 9 \qquad\qquad 6x + 8y = 9$$

Now if you were to subtract the bottom equation from the top equation, the result is $0 = 1$. This statement is *never true*. When this occurs, the system of equations has no solution.

In Examples 1–4, only one equation was multiplied by a number to get the numbers in front of a letter to be the same or opposite. Sometimes each equation must be multiplied by different numbers to get the numbers in front of a letter to be the same or opposite.

Example 5. Solve for x and y.

$$3x + 4y = 5$$
$$5x - 6y = 2$$

Notice that there is no simple number to multiply either equation with to get the numbers in front of x or y to become the same or opposites. In this case, do the following:

1. Select a letter to eliminate.

2. Use the two numbers to the left of this letter. Find the least common multiple of this value as the desired number to be in front of each letter.

3. Determine what value each equation needs to be multiplied by to obtain this value and multiply the equation by that number.

Suppose you want to eliminate x. The least common multiple of 3 and 5, the number in front of the x, is 15. The first equation must be multiplied by 5 in order to get 15 in front of x. The second equation must be multiplied by 3 in order to get 15 in front of x.

$$5(3x) + 5(4y) = 5(5) \qquad 15x + 20y = 25$$
$$3(5x) - 3(6y) = 3(2) \qquad 15x - 18y = 6$$

Now subtract the second equation from the first equation to get the following:

$$38y = 19$$
$$y = \frac{19}{38} \text{ or } y = \frac{1}{2}$$

At this point, you can either replace y with $\frac{1}{2}$ and solve for x (method 1 that follows), or start with the original two equations and eliminate y in order to solve for x (method 2 that follows).

Method 1

Using the top equation: Replace y with $\frac{1}{2}$ and solve for x.

$$3x + 4y = 5$$
$$3x + 4\left(\frac{1}{2}\right) = 5$$
$$3x + 2 = 5$$
$$3x = 3$$
$$x = 1$$

Method 2

Eliminate y and solve for x.

The least common multiple of 4 and 6 is 12. Multiply the top equation by 3 and the bottom equation by 2.

$$3(3x) + 3(4y) = 3(5) \qquad 9x + 12y = 15$$
$$2(5x) - 2(6y) = 2(2) \qquad 10x - 12y = 4$$

Now add the two equations to eliminate y.

$$19x = 19$$
$$x = 1$$

The solution is $x = 1$ and $y = \dfrac{1}{2}$.

Substitution method

Sometimes a system is more easily solved by the *substitution method*. This method involves substituting one equation into another.

Example 6: Solve for x and y.

$$x = y + 8$$
$$x + 3y = 48$$

From the first equation, substitute $(y + 8)$ for x in the second equation.

$$(y + 8) + 3y = 48$$

Now solve for y. Simplify by combining y's.

$$4y + 8 = 48$$
$$\underline{\quad -8 \quad -8 \quad}$$
$$4y = 40$$
$$\frac{4y}{4} = \frac{40}{4}$$
$$y = 10$$

Now insert y's value, 10, in one of the original equations.

$$x = y + 8$$
$$x = 10 + 8$$
$$x = 18$$

Answer: $y = 10$, $x = 18$

Check the solution.

$$x = y + 8 \qquad\qquad x + 3y = 48$$
$$18 = 10 + 8 \qquad 18 + 3(10) = 48$$
$$18 = 18\checkmark \qquad\quad 18 + 30 = 48$$
$$48 = 48\checkmark$$

Example 7. Solve for x and y using the substitution method.

$$5x + 6y = 14$$
$$y - 4x = -17$$

First find an equation that has either a "1" or " -1" in front of a letter. Solve for that letter in terms of the other letter.

Then proceed as in example 6.

In this example, the bottom equation has a "1" in front of the y.

Solve for y in terms of x.

$$y - 4x = -17$$
$$y = 4x - 17$$

Substitute $4x - 17$ for y in the top equation and then solve for x.

$$5x + 6y = 14$$
$$5x + 6(4x - 17) = 14$$
$$5x + 24x - 102 = 14$$
$$29x - 102 = 14$$
$$29x = 116$$
$$x = \frac{116}{29} = 4$$

Replace x with 4 in the equation $y - 4x = -17$ and solve for y.

$$y - 4x = -17$$
$$y - 4(4) = -17$$
$$y - 16 = -17$$
$$y = -1$$

The solution is $x = 4$, $y = -1$.

Check the solution:

$5x + 6y = 14$	$y - 4x = -17$
$5(4) + 6(-1) = 14$	$-1 - 4(4) = -17$
$20 - 6 = 14$	$-1 - 16 = -17$
$14 = 14\checkmark$	$-17 = -17\checkmark$

Graphing method

Another method of solving equations is by *graphing* each equation on a coordinate graph. The coordinates of the intersection will be the solution to the system. If you are unfamiliar with coordinate graphing, carefully review the chapter on coordinate geometry (see Chapter 9) before attempting this method.

Example 8: Solve the system by graphing.

$$x = 4 + y$$
$$x - 3y = 4$$

First, find three values for x and y that satisfy each equation. (Although only two points are necessary to determine a straight line, finding a third point is a good way of checking.) Following are tables of x and y values:

$x = 4 + y$

x	y
4	0
2	-2
5	1

$x - 3y = 4$

x	y
1	-1
4	0
7	1

Now graph the two lines on the coordinate plane, as shown in Figure 5–1.

Figure 5-1 The graph of lines $x = 4 + y$ and $x - 3y = 4$ indicating the solution.

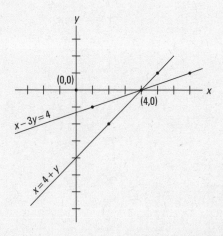

The point where the two lines cross $(4, 0)$ is the solution of the system.

If the lines are parallel, they do not intersect, and therefore, there is no solution to that system.

Example 9. Solve the system by graphing.

$$3x + 4y = 2$$
$$6x + 8y = 4$$

Find three values for x and y that satisfy each equation.

$$3x + 4y = 2 \qquad\qquad 6x + 8y = 4$$

Following are the tables of x and y values. See Figure 5–2.

$3x + 4y = 2$

x	y
0	$\dfrac{1}{2}$
2	-1
4	$-\dfrac{5}{2}$

$6x + 8y = 4$

x	y
0	$\dfrac{1}{2}$
2	-1
4	$-\dfrac{5}{2}$

Figure 5-2   The graph of lines $3x + 4y = 2$ and $6x + 8y = 4$ indicating the solution.

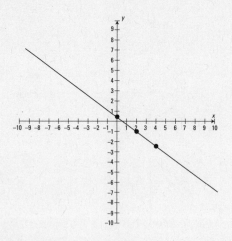

Notice that the same points satisfy each equation. These equations represent the same line.

Therefore, the solution is not a unique point. The solution is all the points on the line.

Therefore, the solution is either equation of the line since they both represent the same line.

This is like Example 3 when it was done using the addition/subtraction method.

Example 10. Solve the system by graphing.

$$3x + 4y = 4$$

$$6x + 8y = 16$$

Find three values for x and y that satisfy each equation. See the following tables of x and y values:

$3x + 4y = 4$		$6x + 8y = 16$	
x	**y**	**x**	**y**
0	1	0	2
2	$-\dfrac{1}{2}$	2	$\dfrac{1}{2}$
4	-2	4	-1

In Figure 5-3, notice that the two graphs are parallel. They will never meet. Therefore, there is no solution for this system of equations.

Figure 5-3 The graph of lines $3x + 4y = 4$ and $6x + 8y = 16$, indicating the solution.

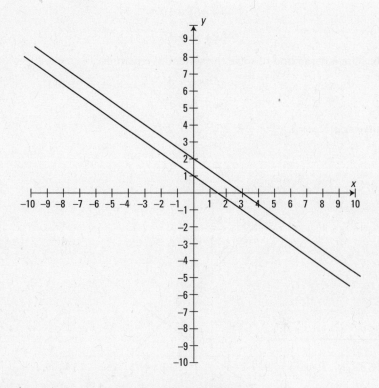

No solution exists for this system of equations.

This is like Example 4 done using the addition/subtraction method.

Chapter Check-Out

1. Use the addition/subtraction method to solve for x and y.

$$8x + 2y = 7$$

$$3x - 4y = 5$$

2. Use the substitution method to solve for a and b.

$$a = b + 1$$

$$a + 2b = 7$$

3. Use the graphing method to solve for x and y.

$$2x - 3y = 0$$

$$x + y = 5$$

4. Use any method to solve the system of equations.

$$4x + 6y = 10$$

$$6x + 9y = 15$$

5. Use any method to solve the system of equations.

$$3x + y = 3$$

$$6x + 2y = 12$$

Answers: 1. $x = 1, y = -\dfrac{1}{2}$ **2.** $a = 3, b = 2$

3.

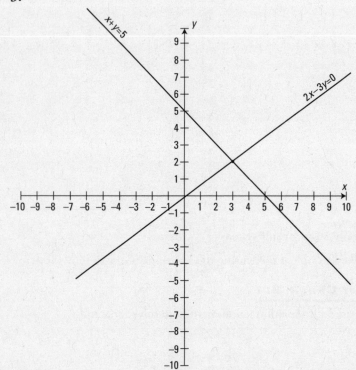

$x = 3, y = 2$

4. No unique solution. Any x and y that makes $4x + 6y = 10$ true or makes $6x + 9y = 15$ true is a solution. Also, any x and y that makes $2x + 3y = 5$ true is also a solution. Both original equations can be simplified to $2x + 3y = 5$. **5.** No solution.

CHAPTER 6

MONOMIALS, POLYNOMIALS, AND FACTORING

Chapter Check-In

❑ Monomials

❑ Operations with monomials

❑ Negative exponents

❑ Polynomials

❑ Operations with polynomials

❑ Factoring

One of the basic skills used in algebra is the ability to work with monomials and polynomials. Factoring polynomial expressions is another important basic skill, but before you can factor, you must review the basics.

Monomials

A **monomial** is an algebraic expression that consists of only one term. (A *term* is a numerical or literal expression with its own sign.) For instance, $9x$, $4a^2$, and $3mpx^2$ are all monomials. The number in front of the variable is called the numerical coefficient. In $9x$, 9 is the coefficient.

Adding and subtracting monomials

To *add* or *subtract monomials,* follow the same rules as with signed numbers (see Chapter 2), *provided that the terms are alike.* Notice that you add or subtract the coefficients only and leave the variables the same.

Example 1: Perform the operation indicated.

(a) $15x^2yz$
$-18x^2yz$
$-3x^2yz$

(b) $3x + 2x = 5x$

(c) $9y$
$-3y$
$6y$

(d) $17q + 8q - 3q - (-4q) =$
$22q - (-4q) =$
$22q + 4q = 26q$

Remember that the rules for signed numbers apply to monomials as well.

Multiplying monomials

Reminder: The rules and definitions for powers and exponents (see Chapter 1) also apply in algebra.

$$5 \cdot 5 = 5^2 \quad \text{and} \quad x \cdot x = x^2$$

Similarly, $a \cdot a \cdot a \cdot b \cdot b = a^3b^2$.

To *multiply monomials,* add the exponents of the same bases.

Example 2: Multiply the following.

(a) $(x^3)(x^4) = x^{3+4} = x^7$

(b) $(x^2y)(x^3y^2) = (x^2x^3)(yy^2) = x^{2+3}y^{1+2} = x^5y^3$

(c) $(6k^5)(5k^2) = (6 \times 5)(k^5k^2) = 30k^{5+2} = 30k^7$ (multiply numbers)

(d) $-4(m^2n)(-3m^4n^3) = [(-4)(-3)](m^2m^4)(nn^3) =$
$12m^{2+4}n^{1+3} = 12m^6n^4$ (multiply numbers)

(e) $(c^2)(c^3)(c^4) = c^{2+3+4} = c^9$

(f) $(3a^2b^3c)(b^2c^2d) = 3(a^2)(b^3b^2)(cc^2)(d) = 3a^2b^{3+2}c^{1+2}d = 3a^2b^5c^3d$

Note that in example (d) the product of -4 and -3 is $+12$, the product of m^2 and m^4 is m^6, and the product of n and n^3 is n^4, because any monomial having no exponent indicated is assumed to have an exponent of l.

When monomials are being raised to a power, the answer is obtained by multiplying the exponents of each part of the monomial by the power to which it is being raised.

Example 3: Simplify.

(a) $(a^7)^3 = a^{21}$

(b) $(x^3y^2)^4 = x^{12}y^8$

(c) $(2x^2y^3)^3 = (2)^3\, x^6y^9 = 8x^6y^9$

Dividing monomials

To *divide monomials,* subtract the exponent of the divisor from the exponent of the dividend of the same base.

Example 4: Divide.

(a) $\dfrac{y^{15}}{y^4} = y^{11}$ or $y^{15} \div y^4 = y^{11}$

(b) $\dfrac{x^5y^2}{x^3y} = x^2y$

(c) $\dfrac{36a^4b^6}{-9ab} = -4a^3b^5$ (divide the numbers)

(d) $\dfrac{fg^{15}}{g^3} = fg^{12}$

(e) $\dfrac{x^5}{x^8} = \dfrac{1}{x^3}$ $\left(\text{may also be expressed as } x^{-3}\right)$

(f) $\dfrac{-3(xy)\left(xy^2\right)}{xy}$

You can simplify the numerator first.

$$\frac{-3(xy)\left(xy^2\right)}{xy} = \frac{-3x^2y^3}{xy} = -3xy^2$$

Or, because the numerator is all multiplication, you can reduce,

$$\frac{-3\left(\overset{1}{\cancel{xy}}\right)\left(xy^2\right)}{\underset{1}{\cancel{xy}}} = -3xy^2$$

Working with negative exponents

Remember, if the exponent is negative, such as x^{-3}, then the variable and exponent may be dropped under the number 1 in a fraction to remove the negative sign as follows.

$$x^{-3} = \frac{1}{x^3}$$

Example 5: Express the answers with positive exponents.

(a) $a^{-2}b = \dfrac{b}{a^2}$

(b) $\dfrac{a^{-3}}{b^4} = \dfrac{1}{a^3 b^4}$

(c) $\left(a^2 b^{-3}\right)\left(a^{-1} b^4\right) = ab$

$$\begin{bmatrix} a^2 \cdot a^{-1} = a \\ b^{-3} \cdot b^4 = b \end{bmatrix}$$

Polynomials

A **polynomial** consists of two or more terms. For example, $x + y$, $y^2 - x^2$, and $x^2 + 3x + 5y^2$ are all polynomials. A **binomial** is a polynomial that consists of exactly two terms. For example, $x + y$ is a binomial. A **trinomial** is a polynomial that consists of exactly three terms. For example, $y^2 + 9y + 8$ is a trinomial.

Polynomials usually are arranged in one of two ways. **Ascending order** is basically when the power of a term increases for each succeeding term. For example, $x + x^2 + x^3$ or $5x + 2x^2 - 3x^3 + x^5$ are arranged in ascending order. **Descending order** is basically when the power of a term decreases for each succeeding term. For example, $x^3 + x^2 + x$ or $2x^4 + 3x^2 + 7x$ are arranged in descending order. Descending order is more commonly used.

Adding and subtracting polynomials

To *add* or *subtract polynomials*, just arrange *like terms* in columns and then add or subtract. (Or simply add or subtract like terms when rearrangement is not necessary.)

Example 6: Do the indicated arithmetic.

(a) Add the polynomials.

$$
\begin{array}{r}
a^2 + ab + b^2 \\
3a^2 + 4ab - 2b^2 \\
\hline
4a^2 + 5ab - b^2
\end{array}
$$

(b) $(5y - 3x) + (9y + 4x) =$
$(5y - 3x) + (9y + 4x) = (5y + 9y) + (-3x + 4x) = 14y + x$ or $x + 14y$

(c) Subtract the polynomials.

$$
\begin{array}{ccc}
a^2 + b^2 & & a^2 + b^2 \\
\underline{-(2a^2 - b^2)} & \rightarrow & \underline{-2a^2 + b^2} \\
& & -a^2 + 2b^2
\end{array}
$$

(d) $(3cd - 6mt) - (2cd - 4mt) =$
$(3cd - 6mt) + (-2cd + 4mt) = (3cd - 2cd) + (-6mt + 4mt) = cd - 2mt$

(e) $3a^2bc + 2ab^2c + 4a^2bc + 5ab^2c =$

$$
\begin{array}{r}
3a^2bc + 2ab^2c \\
+ 4a^2bc + 5ab^2c \\
\hline
7a^2bc + 7ab^2c
\end{array}
$$

or

$$
3a^2bc + 2ab^2c + 4a^2bc + 5ab^2c = \left(3a^2bc + 4a^2bc\right) + \left(2ab^2c + 5ab^2c\right)
$$
$$
= 7a^2bc + 7ab^2c
$$

Multiplying polynomials

To *multiply polynomials,* multiply each term in one polynomial by each term in the other polynomial. Then simplify if necessary.

Example 7: Multiply.

$$
\begin{array}{r}
2x - 2a \\
\times \quad 3x + a \\
\hline
+ 2ax - 2a^2 \\
6x^2 - 6ax \\
\hline
6x^2 - 4ax - 2a^2
\end{array}
\qquad \text{similar to} \qquad
\begin{array}{r}
21 \\
\times \ 23 \\
\hline
63 \\
42 \\
\hline
483
\end{array}
$$

Or you may want to use the "**F.O.I.L.**" method with *binomials*. **F.O.I.L.** means **F**irst terms, **O**utside terms, **I**nside terms, **L**ast terms. Then simplify if necessary.

Example 8: Multiply.

$$(3x + a)(2x - 2a) =$$

Multiply *first* terms from each quantity.

$$\downarrow \ \text{first} \ \downarrow$$
$$(3x + a)(2x - 2a) = 6x^2 \ \rule{2cm}{0.4pt}$$

Now *outside* terms.

$$\downarrow \quad \text{outside} \quad \downarrow$$
$$(3x + a)(2x - 2a) = 6x^2 - 6ax \ \rule{2cm}{0.4pt}$$

Now *inside* terms.

$$\downarrow \ \text{inside} \ \downarrow$$
$$\left(3x + a\right)\left(2x - 2a\right) = 6x^2 - 6ax + 2ax \ \rule{2cm}{0.4pt}$$

Finally *last* terms.

$$\downarrow \quad \text{last} \quad \downarrow$$
$$(3x + a)(2x - 2a) = 6x^2 - 6ax + 2ax - 2a^2$$

Now simplify.

$$6x^2 - 6ax + 2ax - 2a^2 = 6x^2 - 4ax - 2a^2$$

Example 9: Multiply.

$$(x+y)(x+y+z) =$$

$$
\begin{array}{r}
x+y+z \\
\times \quad x+y \\
\hline
xy+y^2+yz \\
+\ x^2+xz+xy \\
\hline
x^2+xz+2xy+y^2+yz
\end{array}
$$

This operation also can be done using the distributive property.

$$(x+y)(x+y+z) = x(x+y+z) + y(x+y+z)$$

$$= x^2 + xy + xz + xy + y^2 + yz$$

$$= x^2 + 2xy + xz + yz + y^2$$

Dividing polynomials by monomials

To *divide a polynomial by a monomial,* just divide each term in the polynomial by the monomial.

Example 10: Divide.

(a) $\left(6x^2 + 2x\right) \div (2x) =$

$$\frac{6x^2 + 2x}{2x} =$$

$$\frac{6x^2}{2x} + \frac{2x}{2x} = 3x + 1$$

(b) $\left(16a^7 - 12a^5\right) \div \left(4a^2\right) =$

$$\frac{16a^7 - 12a^5}{4a^2} =$$

$$\frac{16a^7}{4a^2} - \frac{12a^5}{4a^2} = 4a^5 - 3a^3$$

Dividing polynomials by polynomials

To *divide a polynomial by a polynomial*, make sure both are in descending order; then use long division. (*Remember:* Divide by the first term, multiply, subtract, bring down.)

Example 11: Divide $4a^2 + 18a + 8$ by $a + 4$.

First divide a into $4a^2$

$$
\begin{array}{r}
4a \\
a+4\overline{)4a^2+18a+8}
\end{array}
$$

Now multiply $4a$ times $(a+4)$

$$
\begin{array}{r}
4a \\
a+4\overline{)4a^2+18a+8} \\
4a^2+16a
\end{array}
$$

Now subtract.

$$
\begin{array}{r}
4a \\
a+4\overline{)4a^2+18a+8} \\
-\left(4a^2+16a\right) \\
\hline
2a
\end{array}
$$

Now bring down the 8.

$$
\begin{array}{r}
4a \\
a+4\overline{)4a^2+18a+8} \\
-\left(4a^2+16a\right) \\
\hline
2a+8
\end{array}
$$

Now divide a into $2a$.

$$
\begin{array}{r}
4a \; + \; 2 \\
a+4\overline{)4a^2+18a+8} \\
-\left(4a^2+16a\right) \\
\hline
2a+8
\end{array}
$$

Now multiply 2 times $(a+4)$.

$$\begin{array}{r} 4a+2 \\ a+4\overline{)4a^2+18a+8} \\ -\left(4a^2+16a\right) \\ \hline 2a+8 \\ 2a+8 \end{array}$$

Now subtract.

$$\begin{array}{r} 4a+2 \\ a+4\overline{)4a^2+18a+8} \\ -\left(4a^2+16a\right) \\ \hline 2a+8 \\ -(2a+8) \\ \hline 0 \end{array}$$

$$\begin{array}{r} 4a+2 \\ a+4\overline{)4a^2+18a+8} \\ -\left(4a^2+16a\right) \\ \hline 2a+8 \\ -(2a+8) \\ \hline 0 \end{array}$$

similar to

$$\begin{array}{r} 23 \\ 53\overline{)1219} \\ -(106) \\ \hline 159 \\ -(159) \\ \hline 0 \end{array}$$

Example 12: Divide.

(a) $\left(3x^2+4x+1\right) \div (x+1)$

$$\begin{array}{r} 3x+1 \\ x+1\overline{)3x^2+4x+1} \\ -(3x^2+3x) \\ \hline x+1 \\ -(x+1) \\ \hline 0 \end{array}$$

(b) $\left(2x + 1 + x^2\right) \div (x + 1) =$

First change to descending order: $x^2 + 2x + 1$. Then divide.

$$
\begin{array}{r}
x + 1 \\
x + 1 \overline{)\, x^2 + 2x + 1} \\
\underline{-(x^2 + x)} \\
x + 1 \\
\underline{-(x + 1)} \\
0
\end{array}
$$

(c) $\left(m^3 - m\right) \div (m + 1) =$

Note: When terms are missing, be sure to leave proper room between terms.

$$
\begin{array}{r}
m^2 - m \\
m + 1 \overline{)\, m^3 + 0m^2 - m} \\
\underline{-(m^3 + m^2)} \\
-m^2 - m \\
\underline{-(-m^2 - m)} \\
0
\end{array}
$$

(d) $\left(10a^2 - 29a - 21\right) \div (2a - 7) =$

$$
\begin{array}{r}
5a + 3 \\
2a - 7 \overline{)\, 10a^2 - 29a - 21} \\
\underline{-(10a^2 - 35a)} \\
6a - 21 \\
\underline{-(6a - 21)} \\
0
\end{array}
$$

(e) $\left(x^2 + 2x + 4\right) \div (x + 1) =$

$$x + 1 \text{ (with remainder 3)}$$

$$
\begin{array}{r}
x+1 \overline{\smash{\big)}\, x^2 + 2x + 4} \\
\underline{-(x^2 + x)} \\
x + 4 \\
\underline{-(x + 1)} \\
3
\end{array}
$$

This answer can be rewritten as $\quad x + 1 + \dfrac{3}{x + 1}$

Factoring

To **factor** means to find two or more quantities whose product equals the original quantity.

Factoring out a common factor

To *factor out a common factor,* (1) find the largest common monomial factor of each term and (2) divide the original polynomial by this factor to obtain the second factor. The second factor will be a polynomial.

Example 13: Factor.

(a) $5x^2 + 4x = x(5x + 4)$

(b) $2y^3 - 6y = 2y(y^2 - 3)$

(c) $x^5 - 4x^3 + x^2 = x^2(x^3 - 4x + 1)$

When the common monomial factor is the last term, 1 is used as a place holder in the second factor.

Factoring the difference between two squares

To factor the difference between two squares, (1) find the square root of the first term and the square root of the second term and (2) express your answer as the product of the sum of the quantities from Step 1 times the difference of those quantities.

Example 14: Factor.

(a) $x^2 - 144 = (x + 12)(x - 12)$

   Note: $x^2 + 144$ is not factorable.

(b) $a^2 - b^2 = (a + b)(a - b)$

(c) $9y^2 - 1 = (3y + 1)(3y - 1)$

Factoring polynomials having three terms of the form $ax^2 + bx + c$

To factor polynomials having three terms of the form $ax^2 + bx + c$, (1) check to see whether you can monomial factor (factor out common terms). Then if $a = 1$ (that is, the first term is simply x^2), use double parentheses and factor the first term. Place these factors in the left sides of the parentheses. For example,

$$(x \quad)(x \quad)$$

(2) Factor the last term and place the factors in the right sides of the parentheses.

To decide on the signs of the numbers, do the following. If the sign of the last term is *negative*, (1) find two numbers (one will be a positive number and the other a negative number) whose product is the last term and whose *difference is* the *coefficient* (number in front) of the middle term and (2) give the larger of these two numbers the sign of the middle term and the *opposite* sign to the other factor.

If the sign of the last term is *positive*, (1) find two numbers (both will be positive or both will be negative) whose product is the last term and whose sum is the coefficient of the middle term and (2) give both factors the sign of the middle term.

Example 15: Factor $x^2 - 3x - 10$.

First check to see whether you can monomial factor (factor out common terms). Because this is not possible, use double parentheses and factor the first term as follows: $(x \quad) (x \quad)$. Next, factor the last term, 10, into 2 times 5 (5 must take the negative sign and 2 must take the positive sign because they will then total the coefficient of the middle term, which is -3) and add the proper signs, leaving

$$(x - 5)(x + 2)$$

Multiply **means** (inner terms) and **extremes** (outer terms) to check.

$$\text{outer} \atop 2x$$

$$(x - 5)(x + 2) \quad 2x - 5x = -3x \text{ (which is the middle term)}$$

$$\underset{inner}{\underset{-5x}{}}$$

To completely check, multiply the factors together.

$$
\begin{array}{r}
x - 5 \\
\times \quad x + 2 \\
\hline
+2x - 10 \\
+x^2 - 5x \\
\hline
x^2 - 3x - 10
\end{array}
$$

Example 16: Factor $x^2 + 8x + 15$.

$$(x + 3)(x + 5)$$

Notice that $3 \times 5 = 15$ and $3 + 5 = 8$, the coefficient of the middle term. Also note that the signs of both factors are +, the sign of the middle term. To check,

$$\text{outer} \atop 5x$$

$$(x + 3)(x + 5) \quad 5x + 3x = 8x \text{ (which is the middle term)}$$

$$\underset{inner}{\underset{3x}{}}$$

Example 17: Factor $x^2 - 5x - 14$.

$$(x - 7)(x + 2)$$

Notice that $7 \times 2 = 14$ and $7 - 2 = 5$, the coefficient of the middle term. Also note that the sign of the larger factor, 7, is –, while the other factor, 2, has a + sign. To check,

$$
\overset{\textit{outer}}{\underset{2x}{}}
$$

$$
\downarrow \qquad \downarrow
$$

$$
(x - 7)(x + 2) \qquad 2x - 7x = -5x \text{ (which is the middle term)}
$$

$$
\uparrow \ \uparrow
$$

$$
\underset{\textit{inner}}{-7x}
$$

If, however, $a \neq 1$ (that is, the first term has a coefficient—for example, $4x^2 + 5x + 1$), then additional trial and error will be necessary.

Example 18: Factor $4x^2 + 5x + 1$.

$(2x +)(2x +)$ might work for the first term. But when 1s are used as factors to get the last term, $(2x + 1)(2x + 1)$, the middle term comes out as $4x$ instead of $5x$.

$$
\overset{\textit{outer}}{\underset{2x}{}}
$$

$$
\downarrow \qquad \downarrow
$$

$$
(2x + 1)(2x + 1) \qquad 2x + 2x = 4x
$$

$$
\uparrow \ \uparrow
$$

$$
\underset{\textit{inner}}{2x}
$$

Therefore, try $(4x +)(x +)$. Now using 1s as factors to get the last terms gives $(4x + 1)(x + 1)$. Checking for the middle term,

$$
\overset{\textit{outer}}{\underset{4x}{}}
$$

$$
\downarrow \qquad \downarrow
$$

$$
(4x + 1)(x + 1) \qquad 4x + x = 5x
$$

$$
\uparrow \ \uparrow
$$

$$
\underset{\textit{inner}}{x}
$$

Therefore, $4x^2 + 5x + 1 = (4x + 1)(x + 1)$.

Example 19: Factor $4a^2 + 6a + 2$.

Factoring out a 2 leaves

$$2(2a^2 + 3a + 1)$$

Now factor as usual, giving

$$2(2a + 1)(a + 1)$$

To check,

$$\overset{\substack{outer \\ 2a}}{\downarrow \qquad \downarrow}$$

$(2a + 1)(a + 1)$ $2a + a = 3a$ (the middle term after 2 was factored out)

$$\underset{\substack{a \\ inner}}{\uparrow \ \uparrow}$$

Example 20: Factor $5x^3 + 6x^2 + x$.

Factoring out an x leaves

$$x(5x^2 + 6x + 1)$$

Now factor as usual, giving

$$x(5x + 1)(x + 1)$$

To check,

$$\overset{\substack{outer \\ 5x}}{\downarrow \qquad \downarrow}$$

$(5x + 1)(x + 1)$ $5x + x = 6x$ (the middle term after x was factored out)

$$\underset{\substack{x \\ inner}}{\uparrow \ \uparrow}$$

Example 21: Factor $5 + 7b + 2b^2$ (a slight twist).

$$(5 + 2b)(1 + b)$$

To check,

$$\overset{\overset{\text{\textit{outer}}}{\underset{5b}{}}}{\big\downarrow \qquad \big\downarrow}$$

$$(5 + 2b)(1 + b) \qquad 5b + 2b = 7b \text{ (the middle term)}$$

$$\underset{\underset{\text{\textit{inner}}}{2b}}{\big\uparrow \ \big\uparrow}$$

Note that $(5 + b)(1 + 2b)$ is incorrect because it gives the wrong middle term.

Example 22: Factor $x^2 + 2xy + y^2$.

$$(x + y)(x + y)$$

To check,

$$\overset{\overset{\text{\textit{outer}}}{\underset{xy}{}}}{\big\downarrow \qquad \big\downarrow}$$

$$(x + y)(x + y) \qquad xy + xy = 2xy \text{ (the middle term)}$$

$$\underset{\underset{\text{\textit{inner}}}{xy}}{\big\uparrow \ \big\uparrow}$$

Example 23: Factor $3x^2 - 48$.

Factoring out a 3 leaves

$$3(x^2 - 16)$$

But $x^2 - 16$ is the difference between two squares and can be further factored into $(x + 4)(x - 4)$. Therefore, when completely factored, $3x^2 - 48 = 3(x + 4)(x - 4)$.

Factoring by grouping

Some polynomials have binomial, trinomial, and other polynomial factors.

Example 24: Factor $x + 2 + xy + 2y$.

Since there is no monomial factor, you should attempt rearranging the terms and looking for binomial factors.

$$x + 2 + xy + 2y = x + xy + 2 + 2y$$

Grouping gives

$$(x + xy) + (2 + 2y)$$

Now factoring gives

$$x(1 + y) + 2(1 + y)$$

Using the distributive property gives

$$(x + 2)(1 + y)$$

You could rearrange them differently, but you would still come up with the same factoring.

Summary of the factoring methods

When factoring polynomials, you should look for factoring in the following order.

1. Look for the greatest common factor if one exists.
2. If there are two terms, look for the difference of square numbers.
3. If there are three terms, look for a pattern that applies to $ax^2 + bx + c$.
4. If there are four or more terms, look for some type of regrouping that will lead to other factoring.

Note: There are polynomials that are *not* factorable.

Example 25: Factor $2x^2 + 3x + 5$.

1. This polynomial does not have a common factor.
2. This polynomial is not a difference of square numbers.
3. There is no $(\_x\ \ )(\_x\ \ )$ combination that produces $2x^2 + 3x + 5$.
4. Since there are only three terms, there is no regrouping possibility.

Therefore, this polynomial is not factorable.

Chapter Check-Out

1. $mr^2 + 3mr^2 - 5mr^2 =$
2. $(x^3y^2)(x^2y^3) =$
3. $(4xy^4)^3 =$
4. $\dfrac{x^2 y^4 z^2}{xy^3 z} =$
5. Express the answer with positive exponents: $a^{-3}b =$

6. $(7x + 4y) - (3x - 6y) =$

7. $(2x + y)(3x - y) =$

8. Factor: $6x^3 - 3x^2$

9. Factor: $x^2 - 9$

10. Factor: $x^2 - 4x - 45$

11. Factor: $m^2 - 2mn - 3n^2$

12. Factor: $xy + 3y + x + 3$

Answers: 1. $-mr^2$ **2.** x^5y^5 **3.** $64x^3y^{12}$ **4.** xyz **5.** $\dfrac{b}{a^3}$ **6.** $4x + 10y$ **7.** $6x^2 + xy - y^2$
8. $3x^2(2x - 1)$ **9.** $(x + 3)(x - 3)$ **10.** $(x - 9)(x + 5)$ **11.** $(m - 3n)(m + n)$
12. $(y + 1)(x + 3)$

CHAPTER 7
ALGEBRAIC FRACTIONS

Chapter Check-In

❑ Defining algebraic fractions
❑ Adding algebraic fractions
❑ Subtracting algebraic fractions
❑ Multiplying algebraic fractions
❑ Dividing algebraic fractions

Algebraic fractions are fractions using a variable in the numerator or denominator, such as $\frac{3}{x}$. Because division by 0 is impossible, variables in the denominator have certain restrictions. The denominator can *never* equal 0. Therefore, in the fractions

$$\frac{5}{x} \qquad x \text{ cannot equal } 0 \ (x \neq 0)$$

$$\frac{2}{x-3} \qquad x \text{ cannot equal } 3 \ (x \neq 3)$$

$$\frac{3}{a-b} \qquad \begin{array}{l} a - b \text{ cannot equal } 0 \ (a - b \neq 0) \\ \text{so } a \text{ cannot equal } b \ (a \neq b) \end{array}$$

$$\frac{4}{a^2 b} \qquad \begin{array}{l} \text{neither } a \text{ nor } b \text{ can equal } 0. \\ (a \neq 0 \ , \ b \neq 0) \end{array}$$

Be aware of these types of restrictions.

Operations with Algebraic Fractions

Many techniques will simplify your work as you perform operations with algebraic fractions. As you review the examples in this chapter, note the steps involved in each operation and any methods that will save you time.

Reducing algebraic fractions

To *reduce an algebraic fraction* to lowest terms, first factor the numerator and the denominator; then **reduce**, (or divide out) common factors.

Example 1: Reduce.

(a) $\dfrac{4x^3}{8x^2} = \dfrac{\cancel{4}\,\cancel{x^3}}{\cancel{8}\,\cancel{x^2}} = \dfrac{1}{2}x \text{ or } \dfrac{x}{2}$

(b) $\dfrac{3x-3}{4x-4} = \dfrac{3(x-1)}{4(x-1)} = \dfrac{3\,\cancel{(x-1)}}{4\,\cancel{(x-1)}} = \dfrac{3}{4}$

(c) $\dfrac{x^2+2x+1}{3x+3} = \dfrac{(x+1)(x+1)}{3(x+1)} = \dfrac{\cancel{(x+1)}(x+1)}{3\,\cancel{(x+1)}} = \dfrac{x+1}{3}$

Warning: Do not **reduce** through an addition or subtraction sign as shown here.

$$\frac{x+1}{x+2} \neq \frac{\cancel{x}+1}{\cancel{x}+2} \neq \frac{1}{2}$$

or

$$\frac{x+6}{6} \neq \frac{x+\cancel{6}}{\cancel{6}} \neq x$$

Multiplying algebraic fractions

To *multiply algebraic fractions,* first factor the numerators and denominators that are polynomials; then, reduce where possible. Multiply the remaining numerators together and denominators together. (If you've reduced properly, your answer will be in reduced form.)

Example 2: Multiply.

(a) $\dfrac{2x}{3} \cdot \dfrac{y}{5} = \dfrac{2x \cdot y}{3 \cdot 5} = \dfrac{2xy}{15}$

(b) $\dfrac{x^2}{3y} \cdot \dfrac{2y}{3x} = \dfrac{\overset{1}{\cancel{x^2}}}{3\cancel{y}} \cdot \dfrac{2\cancel{y}}{3\cancel{x}} = \dfrac{2x}{9}$

(c) $\dfrac{x+1}{5y+10} \cdot \dfrac{y+2}{x^2+2x+1} = \dfrac{x+1}{5(y+2)} \cdot \dfrac{y+2}{(x+1)(x+1)}$

$$= \dfrac{\overset{1}{\cancel{(x+1)}}}{5\cancel{(y+2)}} \cdot \dfrac{\overset{1}{\cancel{(y+2)}}}{\cancel{(x+1)}(x+1)}$$

$$= \dfrac{1}{5(x+1)}$$

(d) $\dfrac{x^2-4}{6} \cdot \dfrac{3y}{2x+4} = \dfrac{(x+2)(x-2)}{6} \cdot \dfrac{3y}{2(x+2)}$

$$= \dfrac{\cancel{(x+2)}(x-2)}{\underset{2}{\cancel{6}}} \cdot \dfrac{\overset{1}{\cancel{3}}y}{2\underset{1}{\cancel{(x+2)}}}$$

$$= \dfrac{(x-2)y}{4}$$

(e) $\dfrac{x^2+4x+4}{x-3} \cdot \dfrac{5}{3x+6} = \dfrac{(x+2)(x+2)}{(x-3)} \cdot \dfrac{5}{3(x+2)}$

$$= \dfrac{(x+2)\overset{1}{\cancel{(x+2)}}}{(x-3)} \cdot \dfrac{5}{3\underset{1}{\cancel{(x+2)}}}$$

$$= \dfrac{5(x+2)}{3(x-3)}$$

Dividing algebraic fractions

To *divide algebraic fractions,* invert the **second** fraction and multiply. Remember, you can reduce only after you invert.

Example 3: Divide.

(a) $\dfrac{3x^2}{5} \div \dfrac{2x}{y} = \dfrac{3x^2}{5} \cdot \dfrac{y}{2x} = \dfrac{3x^{\cancel{2}^{1}}}{5} \cdot \dfrac{y}{2\cancel{x}_{1}} = \dfrac{3xy}{10}$

(b) $\dfrac{4x-8}{6} \div \dfrac{x-2}{3} = \dfrac{4x-8}{6} \cdot \dfrac{3}{x-2}$

$$= \dfrac{4(x-2)}{6} \cdot \dfrac{3}{(x-2)}$$

$$= \dfrac{4\,\cancel{(x-2)}^{1}}{\cancel{6}_{2}} \cdot \dfrac{\cancel{3}^{1}}{\cancel{(x-2)}_{1}}$$

$$= \dfrac{4}{2} = 2$$

Adding or subtracting algebraic fractions

To *add* or *subtract algebraic fractions having a common denominator,* simply keep the denominator and combine (add or subtract) the numerators. Reduce if possible.

Example 4: Perform the indicated operation.

(a) $\dfrac{4}{x} + \dfrac{5}{x} = \dfrac{4+5}{x} = \dfrac{9}{x}$

(b) $\dfrac{x-4}{x+1} + \dfrac{3}{x+1} = \dfrac{x-4+3}{x+1} = \dfrac{x-1}{x+1}$

(c) $\dfrac{3x}{y} - \dfrac{2x-1}{y} = \dfrac{3x-(2x-1)}{y} = \dfrac{3x-2x+1}{y} = \dfrac{x+1}{y}$

To *add* or *subtract algebraic fractions having different denominators,* first find a lowest common denominator (LCD), change each fraction to an equivalent fraction with the common denominator, and then combine each numerator. Reduce if possible.

Example 5: Perform the indicated operation.

(a) $\dfrac{2}{x} + \dfrac{3}{y} =$

LCD $= xy$

$$\dfrac{2}{x} \cdot \dfrac{y}{y} + \dfrac{3}{y} \cdot \dfrac{x}{x} = \dfrac{2y}{xy} + \dfrac{3x}{xy} = \dfrac{2y+3x}{xy}$$

(b) $\dfrac{x+2}{3x} + \dfrac{x-3}{6x} =$

LCD $= 6x$

$$\dfrac{x+2}{3x} \cdot \dfrac{2}{2} + \dfrac{x-3}{6x} = \dfrac{2x+4}{6x} + \dfrac{x-3}{6x} =$$

$$\dfrac{2x+4+x-3}{6x} = \dfrac{3x+1}{6x}$$

If there is a common variable factor with more than one exponent, use its greatest exponent.

Example 6: Perform the indicated operation.

(a) $\dfrac{2}{y^2} - \dfrac{3}{y} =$

LCD $= y^2$

$$\dfrac{2}{y^2} - \dfrac{3}{y} \cdot \dfrac{y}{y} = \dfrac{2}{y^2} - \dfrac{3y}{y^2} = \dfrac{2-3y}{y^2}$$

(b) $\dfrac{4}{x^3 y} + \dfrac{3}{xy^2} =$

LCD $= x^3 y^2$

$$\dfrac{4}{x^3 y} \cdot \dfrac{y}{y} + \dfrac{3}{xy^2} \cdot \dfrac{x^2}{x^2} = \dfrac{4y}{x^3 y^2} + \dfrac{3x^2}{x^3 y^2} = \dfrac{4y+3x^2}{x^3 y^2}$$

(c) $\dfrac{x}{x+1} - \dfrac{2x}{x+2} =$

$\text{LCD} = (x+1)(x+2)$

$\dfrac{x}{x+1} \cdot \dfrac{x+2}{x+2} - \dfrac{2x}{x+2} \cdot \dfrac{x+1}{x+1} =$

$\dfrac{x^2 + 2x}{(x+1)(x+2)} - \dfrac{2x^2 + 2x}{(x+1)(x+2)} =$

$\dfrac{x^2 + 2x - 2x^2 - 2x}{(x+1)(x+2)} = \dfrac{-x^2}{(x+1)(x+2)}$

To find the lowest common denominator, it is often necessary to factor the denominators and proceed as follows.

Example 7: Perform the indicated operation.

$\dfrac{2x}{x^2 - 9} - \dfrac{5}{x^2 + 4x + 3} = \dfrac{2x}{(x+3)(x-3)} - \dfrac{5}{(x+3)(x+1)}$

$\text{LCD} = (x+3)(x-3)(x+1)$

$\dfrac{2x}{(x+3)(x-3)} \cdot \dfrac{(x+1)}{(x+1)} - \dfrac{5}{(x+3)(x+1)} \cdot \dfrac{(x-3)}{(x-3)} =$

$\dfrac{2x^2 + 2x}{(x+3)(x-3)(x+1)} - \dfrac{5x - 15}{(x+3)(x-3)(x+1)} =$

$\dfrac{2x^2 + 2x - (5x - 15)}{(x+3)(x-3)(x+1)} = \dfrac{2x^2 + 2x - 5x + 15}{(x+3)(x-3)(x+1)} =$

$\dfrac{2x^2 - 3x + 15}{(x+3)(x-3)(x+1)}$

Occasionally, a problem will require reducing what appears to be the final result. A problem like that is found in the next example.

Example 8: Perform the indicated operation.

$$\frac{x}{x^2+5x+6} - \frac{2}{x^2+3x+2} = \frac{x}{(x+3)(x+2)} - \frac{2}{(x+2)(x+1)}$$

$$\text{LCD} = (x+1)(x+2)(x+3)$$

$$\frac{x}{(x+3)(x+2)} \cdot \frac{(x+1)}{(x+1)} - \frac{2}{(x+2)(x+1)} \cdot \frac{(x+3)}{(x+3)} =$$

$$\frac{x^2+x}{(x+1)(x+2)(x+3)} - \frac{2x+6}{(x+1)(x+2)(x+3)} =$$

$$\frac{x^2+x-2x-6}{(x+1)(x+2)(x+3)} = \frac{x^2-x-6}{(x+1)(x+2)(x+3)} =$$

$$\frac{(x-3)(x+2)}{(x+1)(x+2)(x+3)} = \frac{(x-3)\cancel{(x+2)}^{1}}{(x+1)\cancel{(x+2)}_{1}(x+3)} =$$

$$\frac{x-3}{(x+1)(x+3)}$$

Chapter Check-Out

1. Reduce: $\dfrac{9x^5}{12x^3}$

2. Reduce: $\dfrac{x^2-9x+20}{x^2-x-12}$

3. $\dfrac{x-1}{x} \cdot \dfrac{x^2+3x}{x^2-7x+6} =$

4. $\dfrac{10y+5}{4} \div \dfrac{2y+1}{2} =$

5. $\dfrac{7}{x} + \dfrac{3}{y} =$

6. $\dfrac{3}{x^4y^2} - \dfrac{2}{x^2y^3} =$

7. $\dfrac{2x}{x-1} - \dfrac{x}{x+2} =$

8. $\dfrac{x}{x^2+11x+30} - \dfrac{5}{x^2+9x+20}$

Answers: 1. $\dfrac{3x^2}{4}$ **2.** $\dfrac{x-5}{x+3}$ **3.** $\dfrac{x+3}{x-6}$ **4.** $\dfrac{5}{2}$ **5.** $\dfrac{7y+3x}{xy}$ or $\dfrac{3x+7y}{xy}$

6. $\dfrac{3y-2x^2}{x^4y^3}$ **7.** $\dfrac{x^2+5x}{x^2+x-2}$ **8.** $\dfrac{x-6}{(x+4)(x+6)}$

CHAPTER 8

INEQUALITIES, GRAPHING, AND ABSOLUTE VALUE

Chapter Check-In

❑ Inequalities and their properties

❑ Solving and graphing inequalities

❑ Absolute value

❑ Solving absolute value equations

❑ Solving inequalities containing absolute values

Having reviewed solving equations and working with monomials, you are now ready to work with inequalities.

Inequalities

An **inequality** is a statement in which the relationships are not equal. Instead of using an equal sign (=) as in an equation, these symbols are used: > (is greater than) and < (is less than) or ≥ (is greater than or equal to) and ≤ (is less than or equal to).

Axioms and properties of inequalities

For all real numbers a, b, and c, the following are some basic rules for using the inequality signs.

■ **Trichotomy axiom:** $a > b$, $a = b$, or $a < b$.

These are the only possible relationships between two numbers. Either the first number is greater than the second, the numbers are equal, or the first number is less than the second.

- **Transitive axiom:** If $a > b$, and $b > c$, then $a > c$.

    Therefore, if $3 > 2$ and $2 > 1$, then $3 > 1$.

    If $a < b$ and $b < c$, then $a < c$.

    Therefore, if $4 < 5$ and $5 < 6$, then $4 < 6$.

- **Addition property:**

    If $a > b$, then $a + c > b + c$.

    If $a > b$, then $a - c > b - c$.

    If $a < b$, then $a + c < b + c$.

    If $a < b$, then $a - c < b - c$.

    Adding or subtracting the same amount from each side of an inequality keeps the direction of the inequality the same.

    Example: If $3 > 2$, then $3 + 1 > 2 + 1$ $(4 > 3)$

    If $12 < 15$, then $12 - 4 < 15 - 4$ $(8 < 11)$

- **Positive multiplication and division property:**

    If $a > b$, and $c > 0$, then $ac > bc$.

    If $a < b$, and $c > 0$, then $ac < bc$.

    If $a > b$, and $c > 0$, then $\dfrac{a}{c} > \dfrac{b}{c}$.

    If $a < b$, and $c > 0$, then $\dfrac{a}{c} < \dfrac{b}{c}$.

    Multiplying or dividing each side of an inequality by a positive number keeps the direction of the inequality the same.

    Example: If $5 > 2$, then $5(3) > 2(3)$, therefore, $15 > 6$.

    If $3 < 12$, then $\dfrac{3}{4} < \dfrac{12}{4}$ $\left(\dfrac{3}{4} < 3 \right)$.

- **Negative multiplication and division property:**

    If $a > b$, and $c < 0$, then $ac < bc$.

    If $a < b$, and $c < 0$, then $ac > bc$.

If $a > b$, and $c < 0$, then $\dfrac{a}{c} < \dfrac{b}{c}$.

If $a < b$, and $c < 0$, then $\dfrac{a}{c} > \dfrac{b}{c}$.

Multiplying or dividing each side of an inequality by a negative number reverses the direction of the inequality.

Example: If $5 > 2$, then $5(-3) < 2(-3)$; therefore, $-15 < -6$.

If $3 < 12$, then $\dfrac{3}{-4} > \dfrac{12}{-4}$ $\left(-\dfrac{3}{4} > -3\right)$.

Solving inequalities

When working with inequalities, treat them exactly like equations (except, if you multiply or divide each side of the inequality by a negative number, you must reverse the direction of the inequality).

Example 1: Solve for x: $2x + 4 > 6$.

$$2x + 4 > 6$$
$$\underline{-4 \quad -4}$$
$$2x \quad > 2$$
$$\dfrac{2x}{2} > \dfrac{2}{2}$$
$$x > 1$$

Answers are sometimes written in **set builder notation** $\{x: x > 1\}$, which is read "the set of all x such that x is greater than 1."

Example 2: Solve for x: $-7x > 14$.

Divide by -7 and reverse the direction of the inequality.

$$\dfrac{-7x}{-7} < \dfrac{14}{-7}$$
$$x < -2$$

Example 3: Solve for x: $3x + 2 \geq 5x - 10$.

$$3x + 2 \geq 5x - 10$$
$$\underline{\quad -2 \qquad -2 \quad}$$
$$3x \quad \geq 5x - 12$$

$$3x \quad \geq \quad 5x - 12$$
$$\underline{-5x \qquad -5x \quad}$$
$$-2x \quad \geq \quad -12$$

$$\frac{-2x}{-2} \leq \frac{-12}{-2}$$
$$x \leq 6$$

Notice that opposite operations are used. Divide each side of the inequality by -2 and reverse the direction of the inequality.

In set builder notation, $\{x: x \leq 6\}$.

Graphing on a Number Line

Integers and real numbers can be represented on a **number line.** The point on this line associated with each number is called the *graph* of the number. Notice that number lines are spaced equally, or proportionately (see Figure 8–1).

Figure 8–1 Number lines.

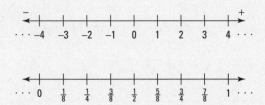

Graphing inequalities

When graphing inequalities involving only integers, dots are used.

Example 4: Graph the set of x such that $1 \leq x \leq 4$ and x is an integer (see Figure 8–2).

$$\{x : 1 \leq x \leq 4, x \text{ is an integer}\}$$

Figure 8–2 A graph of $\{x : 1 \leq x \leq 4, x \text{ is an integer}\}$.

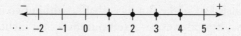

When *graphing inequalities involving real numbers,* lines, rays, and dots are used. A dot is used if the number is included. A hollow dot is used if the number is not included.

Example 5: Graph as indicated (see Figure 8–3).

(a) Graph the set of x such that $x \geq 1$.

$$\{x : x \geq 1\}$$

Figure 8–3 A graph of $\{x : x \geq 1\}$.

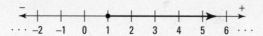

(b) Graph the set of x such that $x > 1$ (see Figure 8–4).

$$\{x : x > 1\}$$

Figure 8–4 A graph of $\{x : x > 1\}$

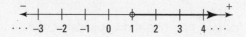

(c) Graph the set of x such that $x < 4$ (see Figure 8–5).

$$\{x : x < 4\}$$

Figure 8–5 A graph of {x: x < 4}

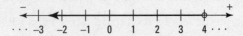

This ray is often called an **open ray** or a *half line*. The **hollow dot** distinguishes an open ray from a ray.

Intervals

An **interval** consists of all the numbers that lie within two certain boundaries. If the two boundaries, or fixed numbers, are included, then the interval is called a **closed interval**. If the fixed numbers are not included, then the interval is called an **open interval**.

Example 6: Graph.

(a) Closed interval (see Figure 8–6).

$$\{x: -1 \le x \le 2\}$$

Figure 8–6 A graph showing closed interval {x: −1 ≤ x ≤ 2}.

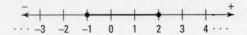

(b) Open interval (see Figure 8–7).

$$\{x: -2 < x < 2\}$$

Figure 8–7 A graph showing open interval {x: −2 < x < 2}.

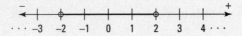

If the interval includes only one of the boundaries, then it is called a **half-open interval**.

Example 7: Graph the half-open interval (see Figure 8–8).

$$\{x: -1 < x \le 2\}$$

Figure 8–8 A graph showing half-open interval {x: $-1 < x \leq 2$}.

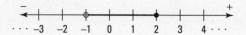

Absolute Value

The numerical value when direction or sign is not considered is called the **absolute value.** The absolute value of x is written $|x|$. The absolute value of a number is always positive except when the number is 0.

$$|0| = 0$$

$$|x| > 0 \text{ when } x \neq 0$$

$$|-x| > 0 \text{ when } x \neq 0$$

Example 8: Give the value.

(a) $|4| = 4$

(b) $|-6| = 6$

(c) $|7 - 9| = |-2| = 2$

(d) $3 - |-6| = 3 - 6 = -3$

Note that absolute value is taken first, or work is first done within the absolute value signs.

Solving equations containing absolute value

To *solve an equation containing absolute value,* isolate the absolute value on one side of the equation. Then set its contents equal to both the positive and negative value of the number on the other side of the equation and solve both equations.

Example 9. Solve $|x| + 2 = 5$.

Isolate the absolute value.

$$\begin{array}{rr} |x| + 2 = & 5 \\ -2 & -2 \\ \hline |x| & = 3 \end{array}$$

Set the contents of the absolute value portion equal to +3 and −3.

$$x = 3 \quad \text{or} \quad x = -3$$

Answer: 3, −3

Example 10: Solve $3|x - 1| - 1 = 11$.

Isolate the absolute value.

$$
\begin{array}{r}
3|x-1|-1=11 \\
+1 \quad +1 \\
\hline
3|x-1| \quad = 12
\end{array}
$$

$$\frac{3|x-1|}{3} = \frac{12}{3}$$
$$|x-1| = 4$$

Set the contents of the absolute value portion equal to +4 and −4.

Solving for x,

$$
\begin{array}{rcl}
x - 1 = 4 & \text{or} & x - 1 = -4 \\
+1 \quad +1 & & +1 \quad +1 \\
\hline
x \quad = 5 & \text{or} & x \quad = -3
\end{array}
$$

Answer: 5, −3

Solving inequalities containing absolute value and graphing

To *solve an inequality containing absolute value*, begin with the same steps as for solving equations with absolute value. When creating the comparisons to both the + and − of the other side of the inequality, reverse the direction of the inequality when comparing with the negative.

Example 11: Solve and graph the answer: $|x - 1| > 2$.

Notice that the absolute value expression is already isolated.

$$|x - 1| > 2$$

Compare the contents of the absolute value portion to both 2 and –2. Be sure to reverse the direction of the inequality when comparing it with –2.

Solve for x.

$$x - 1 > 2 \quad \text{or} \quad x - 1 < -2$$
$$\underline{+1 \quad +1} \qquad\qquad \underline{+1 \quad +1}$$
$$x \quad\quad > 3 \quad \text{or} \quad x \quad\quad < -1$$

Graph the answer (see Figure 8–9).

Figure 8–9 The graphic solution to $|x - 1| > 2$.

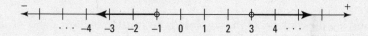

Example 12: Solve and graph the answer: $3|x| - 2 \le 1$.

Isolate the absolute value.

$$3|x| - 2 \le 1$$
$$\underline{+2 \quad +2}$$
$$3|x| \quad \le 3$$
$$\frac{3|x|}{3} \le \frac{3}{3}$$
$$|x| \le 1$$

Compare the contents of the absolute value portion to both 1 and –1. Be sure to reverse the direction of the inequality when comparing it with –1.

$$x \le 1 \qquad \text{or} \qquad x \ge -1$$

Graph the answer (see Figure 8–10).

Figure 8–10 Graphing the solution to $3|x| - 2 \le 1$.

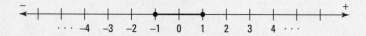

Example 13: Solve and graph the answer: $2|1-x| + 1 \geq 3$.

Isolate the absolute value.

$$2|1-x|+1 \geq 3$$
$$\underline{-1 \quad -1}$$
$$2|1-x| \quad \geq 2$$

$$\frac{2|1-x|}{2} \geq \frac{2}{2}$$
$$|1-x| \geq 1$$

Compare the contents of the absolute value portion to both 1 and –1. Be sure to reverse the direction of the inequality when comparing it with –1.

Solve for x.

$1-x \geq 1$	or	$1-x \leq -1$
$\underline{-1 \quad\quad -1}$		$\underline{-1 \quad\quad -1}$
$-x \geq 0$		$-x \leq -2$

$$\frac{-x}{-1} \leq \frac{0}{-1} \qquad\qquad \frac{-x}{-1} \geq \frac{-2}{-1}$$

(Remember to switch the direction of the inequality when dividing by a negative)

$$x \leq 0 \qquad \text{or} \qquad x \geq 2$$

Graph the answer (see Figure 8–11).

Figure 8–11 Graphing the solution $2|1 -x| + 1 \geq 3$.

Chapter Check-Out

1. True or false: If $a > b$ and $b > c$ then $a > c$.

2. Solve for x: $7x + 3 \leq 9x - 7$

3. Graph: $\{x: x > -2\}$

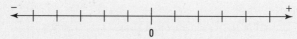

4. Graph: $\{x: 4 > x \geq -2$

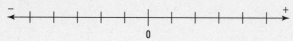

5. $3 - |-5| =$

6. Solve for x: $4|x - 1| - 3 = 17$

7. Solve and graph $2|x| + 4 > 8$.

8. Solve and graph $5|x - 2| + 8 \leq 33$.

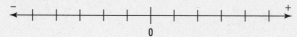

Answers: 1. True **2.** $x \geq 5$

3.

4.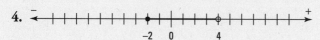

5. -2

6. $-4, 6$

7. $x > 2$ or $x < -2$

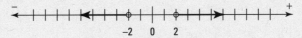

8. $-3 \le x \le 7$

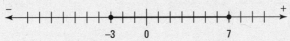

CHAPTER 9
COORDINATE GEOMETRY

Chapter Check-In

❑ Coordinate graphs

❑ Graphing equations on coordinate planes

❑ Slope and intercept of linear equations

❑ Finding equations of lines

❑ Linear inequalities and half-planes

Coordinate geometry deals with graphing (or plotting) and analyzing points, lines, and areas on the coordinate plane (coordinate graph).

Coordinate Graphs

Each point on a number line is assigned a number. In the same way, each point in a plane is assigned a pair of numbers. These numbers represent the placement of the point relative to two intersecting lines. In **coordinate graphs** (see Figure 9–1), two perpendicular number lines are used and are called **coordinate axes.** One axis is horizontal and is called the **x-axis.** The other is vertical and is called the **y-axis.** The point of intersection of the two number lines is called the **origin** and is represented by the coordinates (0, 0).

Figure 9–1 An *x*-*y* coordinate graph.

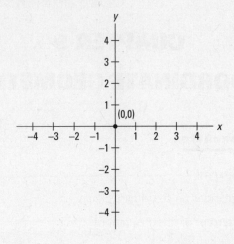

Each point on a plane is located by a unique **ordered pair** of numbers called the *coordinates*. Some coordinates are noted in Figure 9–2.

Figure 9–2 Graphing or plotting coordinates.

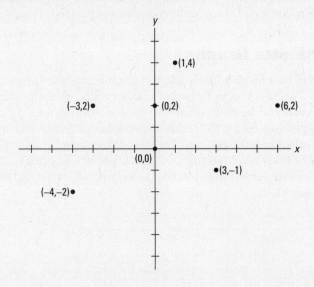

Notice that on the *x*-axis numbers to the right of 0 are positive and to the left of 0 are negative. On the *y*-axis, numbers above 0 are positive and below 0 are negative. Also, note that the first number in the ordered pair is called the *x*-coordinate, or **abscissa,** and the second number is the *y*-coordinate, or **ordinate.** The *x*-coordinate shows the right or left direction, and the *y*-coordinate shows the up or down direction.

The coordinate graph is divided into four quarters called **quadrants.** These quadrants are labeled in Figure 9–3.

Figure 9–3 Coordinate graph with quadrants labeled.

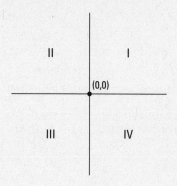

Notice the following:

In quadrant I, *x* is always positive and *y* is always positive.

In quadrant II, *x* is always negative and *y* is always positive.

In quadrant III, *x* and *y* are both always negative.

In quadrant IV, *x* is always positive and *y* is always negative.

Graphing equations on the coordinate plane

To *graph an equation on the coordinate plane,* find the coordinate by giving a value to one variable and solving the resulting equation for the other value. Repeat this process to find other coordinates. (When giving a value for one variable, you could start with 0, then try 1, and so on.) Then graph the solutions.

Example 1: Graph the equation $x + y = 6$.

If $x = 0$, then $y = 6$. $\qquad (0) + y = 6$

$\qquad\qquad\qquad\qquad y = 6$

If $x = 1$, then $y = 5$ $\qquad (1) + y = 6$

$$\dfrac{-1 \qquad -1}{}$$

$$y = 5$$

If $x = 2$, then $y = 4$. $(2) + y = 6$

$$\underline{ -2 \qquad -2}$$

$$y = 4$$

Using a simple chart is helpful.

x	y
0	6
1	5
2	4

Now plot these coordinates as shown in Figure 9–4.

Figure 9–4 Plotting of coordinates (0,6), (1,5), (2,4)

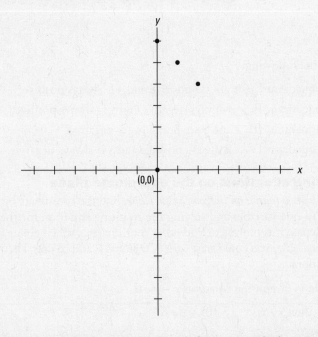

Notice that these solutions, when plotted, form a straight line. Equations whose solution sets form a straight line are called **linear equations**. Complete the graph of $x + y = 6$ by drawing the line that passes through these points (see Figure 9-5).

Figure 9–5   The line that passes through the points graphed in Figure 9–4.

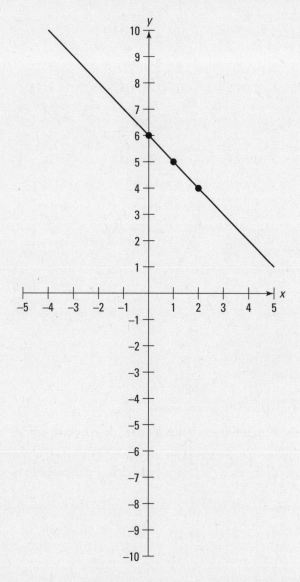

Equations that have a variable raised to a power, show division by a variable, involve variables with square roots, or have variables multiplied together will not form a straight line when their solutions are graphed. These are called **nonlinear equations.**

Example 2: Graph the equation $y = x^2 + 4$.

If $x = 0$, then $y = 4$.
$$y = (0)^2 + 4$$
$$y = 0 + 4$$
$$y = 4$$

If $x = 1$ or -1, then $y = 5$.
$$y = (1)^2 + 4 \qquad y = (-1)^2 + 4$$
$$y = 1 + 4 \qquad y = 1 + 4$$
$$y = 5 \qquad y = 5$$

If $x = 2$ or -2, then $y = 8$.
$$y = (2)^2 + 4 \qquad y = (-2)^2 + 4$$
$$y = 4 + 4 \qquad y = 4 + 4$$
$$y = 8 \qquad y = 8$$

Use a simple chart.

x	y
-2	8
-1	5
0	4
1	5
2	8

Now plot these coordinates as shown in Figure 9–6.

Notice that these solutions, when plotted, do not form a straight line.

These solutions, when plotted, give a curved line (nonlinear). The more points plotted, the easier it is to see and describe the solutions set.

Figure 9–6 Plotting the coordinates in the simple chart.

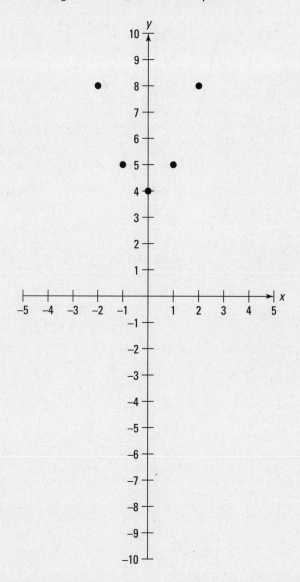

Complete the graph of $y = x^2 + 4$ by connecting these points with a smooth curve that passes through these points (see Figure 9–7).

Figure 9–7 The line that passes through the points graphed in Figure 9–6.

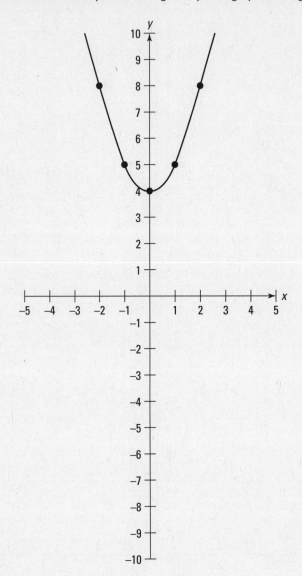

Slope and intercept of linear equations

There are two relationships between the graph of a linear equation and the equation itself that must be pointed out. One involves the *slope of the line,* and the other involves the point where the *line crosses the y-axis.* In order to see either of these relationships, the terms of the equation must be in a certain order.

$$(+)(1)y = (\ )x + (\ )$$

When the terms are written in this order, the equation is said to be in *y*-form. *Y*-form is written **y = mx + b**, and the two relationships involve *m* and *b*.

Example 3: Write the equations in *y*-form.

(a) $x - y = 3$

$$-y = -x + 3$$

$$y = x - 3$$

(b) $y = -2x + 1$ (already in *y*-form)

(c) $x - 2y = 4$

$$-2y = -x + 4$$

$$2y = x - 4$$

$$y = \frac{1}{2}x - 2$$

As shown in the graphs of the three problems in Figure 9–8, the lines cross the *y*-axis at –3, +1, and –2, the last term in each equation.

If a linear equation is written in the form of $y = mx + b$, *b* is the *y*-intercept.

The *slope* of a line is defined as

$$\frac{\text{the change in } y}{\text{the change in } x}$$

and the word "change" refers to the difference in the value of *y* (or *x*) between two points on the line.

$$\text{slope of line } AB = \frac{y_A - y_B}{x_A - x_B} \left[\frac{y \text{ at point } A - y \text{ at point } B}{x \text{ at point } A - x \text{ at point } B} \right]$$

Note: Points *A* and *B* can be any two points on a line; there will be no difference in the slope.

Figure 9–8 Graphs showing the lines crossing the *y*-axis.

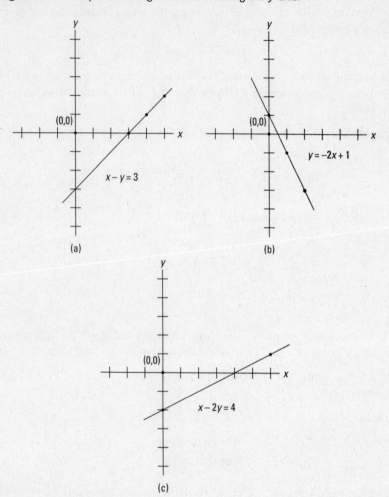

(a)

(b)

(c)

Example 4: Find the slope of $x - y = 3$ using coordinates.

To find the slope of the line, pick any two points on the line, such as $A\,(3, 0)$ and $B\,(5, 2)$, and calculate the slope.

$$\text{slope} = \frac{y_A - y_B}{x_A - x_B} = \frac{(0) - (2)}{(3) - (5)} = \frac{-2}{-2} = 1$$

Example 5: Find the slope of $y = -2x + 1$ using coordinates.

Pick two points, such as $A\,(1, -1)$ and $B\,(2, -3)$, and calculate the slope.

$$\text{slope} = \frac{y_A - y_B}{x_A - x_B} = \frac{-1-(-3)}{1-2} = \frac{-1+3}{1-2} = \frac{2}{-1} = -2$$

Example 6: Find the slope of $x - 2y = 4$ using coordinates.

Pick two points, such as $A\,(0, -2)$ and $B\,(4, 0)$, and calculate the slope.

$$\text{slope} = \frac{y_A - y_B}{x_A - x_B} = \frac{(-2)-(0)}{(0)-(4)} = \frac{-2}{-4} = \frac{1}{2}$$

Looking back at the equations for Example 3(a), (b), and (c) written in y-form, it should be evident that the slope of the line is the same as the numerical coefficient of the x-term.

(a) $y = 1x - 3$

slope = 1   y-intercept = -3

(b) $y = -2x + 1$

slope = -2   y-intercept = 1

(c) $y = \frac{1}{2}x - 2$

slope = $\frac{1}{2}$ y-intercept = -2

Graphing linear equations using slope and intercept

Graphing an equation by using its slope and y-intercept is usually quite easy.

1. State the equation in y-form.
2. Locate the y-intercept on the graph (that is, one of the points on the line).
3. Write the slope as a ratio (fraction) and use it to locate other points on the line.
4. Draw the line through the points.

Example 7: Graph the equation $x - y = 2$ using slope and y-intercept.

$$x - y = 2$$
$$-y = -x + 2$$
$$y = x - 2$$

Locate -2 on the y-axis and from this point count as shown in Figure 9–9:

slope $= 1$

$$\text{or } \frac{1}{1} \quad \begin{matrix} \text{(for every 1 up)} \\ \text{(go 1 to the right)} \end{matrix}$$

$$\text{or } \frac{-1}{-1} \quad \begin{matrix} \text{(for every 1 down)} \\ \text{(go 1 to the left)} \end{matrix}$$

Figure 9–9 Graph of line $y = x - 2$.

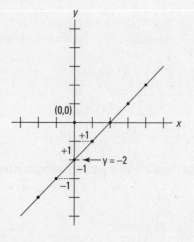

Example 8: Graph the equation $2x - y = -4$ using slope and y-intercept.

$$2x - y = -4$$
$$-y = -2x - 4$$
$$y = 2x + 4$$

Locate +4 on the *y*-axis and from this point count as shown in Figure 9–10:

slope = 2

$$\text{or } \frac{2}{1} \quad \begin{array}{l}\text{(for every 2 up)}\\\text{(go 1 to the right)}\end{array}$$

$$\text{or } \frac{-2}{-1} \quad \begin{array}{l}\text{(for every 2 down)}\\\text{(go 1 to the left)}\end{array}$$

Figure 9–10 Graph of line $2x - y = -4$.

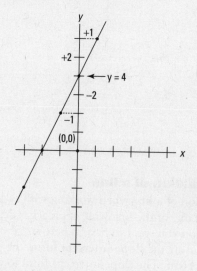

Example 9: Graph the equation $x + 3y = 0$ using slope and *y*-intercept.

$$x + 3y = 0$$

$$3y = -x + (0)$$

$$y = -\frac{1}{3}x + (0)$$

Locate 0 on the y-axis and from this point count as shown in Figure 9–11:
slope $= -\dfrac{1}{3}$

$$\text{or } \dfrac{-1}{3} \begin{array}{l} \text{(for every 1 down)} \\ \text{(go 3 to the right)} \end{array}$$

$$\text{or } \dfrac{1}{-3} \begin{array}{l} \text{(for every 1 up)} \\ \text{(go 3 to the left)} \end{array}$$

Figure 9–11 Graph of line $x + 3y = 0$.

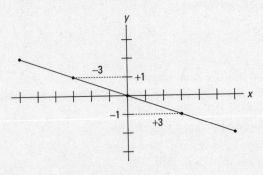

Finding the equation of a line

To find the equation of a line when working with ordered pairs, slopes, and intercepts, use one of the following approaches depending on which form of the equation you want to have. There are several forms, but the three most common are the **slope-intercept form**, the **point-slope form**, and the **standard form**. The slope-intercept form looks like $y = mx + b$ where m is the slope of the line and b is the y-intercept. The point-slope form looks like $y - y_1 = m(x - x_1)$ where m is the slope of the line and (x_1, y_1) is any point on the line. The standard form looks like $Ax + By = C$ where, if possible, A, B, and C are integers.

Method 1. Slope–intercept form.

1. Find the slope, m.
2. Find the y-intercept, b.
3. Substitute the slope and y-intercept into the slope-intercept form, $y = mx + b$.

Method 2. Point-slope form.

1. Find the slope, m.
2. Use any point known to be on the line.
3. Substitute the slope and the ordered pair of the point into the point-slope form, $y - y_1 = m(x - x_1)$.

Note: You could begin with the point-slope form for the equation of the line and then solve the equation for y. You will get the slope-intercept form without having to first find the y-intercept.

Method 3. Standard form.

1. Find the equation of the line using either the slope-intercept form or the point-slope form.
2. With appropriate algebra, arrange to get the x-terms and the y-terms on one side of the equation and the constant on the other side of the equation.
3. If necessary, multiply each side of the equation by the least common denominator of all the denominators to have all integer coefficients for the variables.

Example 10: Find the equation of the line, in slope-intercept form, when $m = -4$ and $b = 3$. Then convert it into standard form.

1. Find the slope, m.

$$m = -4 \text{ (given)}$$

2. Find the y-intercept, b.

$$b = 3 \text{ (given)}$$

3. Substitute the slope and y-intercept into the slope-intercept form, $y = mx + b$.

$$y = -4x + 3 \text{ (slope-intercept form)}$$

4. With appropriate algebra, arrange to get the x-terms and the y-terms on one side of the equation and the constant on the other side of the equation.

$$
\begin{array}{r}
y = -4x + 3 \\
\underline{+4x \qquad +4x} \\
4x + y = \qquad 3 \quad \text{(standard form)}
\end{array}
$$

Example 11: Find the equation of the line, in point-slope form, passing through the point (6, 4) with a slope of −3. Then convert it to standard form.

1. Find the slope, *m*.

$$m = -3 \text{ (given)}$$

2. Use any point known to be on the line.

$$(6, 4) \text{ (given)}$$

3. Substitute the slope and the ordered pair of the point into the point-slope form,

$$y - y_1 = m(x - x_1)$$
$$y - 4 = -3(x - 6) \quad \text{(point-slope form)}$$

4. With appropriate algebra, arrange to get the *x*-terms and the *y*-terms on one side of the equation and the constant on the other side of the equation.

$$y - 4 = -3(x - 6)$$
$$y - 4 = -3x + 18$$

$$
\begin{array}{r}
+3x \qquad\quad +3x \\
\hline
3x + y - 4 = \qquad 18 \\
+4 \qquad +4 \\
\hline
3x + y \quad = \qquad 22 \quad \text{(standard form)}
\end{array}
$$

Example 12: Find the equation of the line, in either slope-intercept form or point-slope form, passing through (5, −4) and (3, 7). Then convert it to standard form.

Starting with slope-intercept:

1. Find the slope, *m*.
$$m = \frac{\text{change in } y}{\text{change in } x}$$

$$m = \frac{(-4) - 7}{5 - 3} = \frac{-11}{2} = -\frac{11}{2}$$

2. Find the y-intercept, b.

    Substitute the slope and either point into the slope-intercept form.

    $$y = mx + b \text{ where } m = -\frac{11}{2}, x = 5, y = -4$$

    $$-4 = -\frac{11}{2}(5) + b$$

    $$-4 = -\frac{55}{2} + b$$

    $$\underline{+\frac{55}{2} \quad +\frac{55}{2}}$$

    $$\frac{47}{2} = \quad b$$

3. Substitute the slope and y-intercept into the slope-intercept form, $y = mx + b$.

    Since $\quad m = -\frac{11}{2}$

    and $\quad b = \frac{47}{2}$

    then $\quad y = -\frac{11}{2}x + \frac{47}{2}$ $\quad$ (slope-intercept form)

4. With appropriate algebra, arrange to get the x-terms and the y-terms on one side of the equation and the constant on the other side of the equation.

    If necessary, multiply each side of the equation by the least common denominator of all the denominators to have all integer coefficients for the variables.

    $$y = -\frac{11}{2}x + \frac{47}{2}$$

    $$\underline{+\frac{11}{2}x \qquad +\frac{11}{2}x}$$

    $$\frac{11}{2}x + y = \qquad \frac{47}{2}$$

    $$2\left(\frac{11}{2}x + y\right) = 2\left(\frac{47}{2}\right)$$

    $$11x + 2y = 47 \qquad \text{(standard form)}$$

Starting with point-slope form:

1. Find the slope, m. $m = \dfrac{\text{change in } y}{\text{change in } x}$

$$m = \frac{(-4) - 7}{5 - 3} = \frac{-11}{2} = -\frac{11}{2}$$

2. Use any point known to be on the line.

$$(3, 7) \quad \text{(given)}$$

3. Substitute the slope and the ordered pair of the point into the point-slope form,

$$y - y_1 = m(x - x_1)$$

$$y - 7 = -\frac{11}{2}(x - 3) \quad \text{(point-slope form)}$$

4. With appropriate algebra, arrange to get the x-terms and the y-terms on one side of the equation and the constant on the other side of the equation.

   If necessary, multiply each side of the equation by the least common denominator to have all integer coefficients for the variables.

$$y - 7 = -\frac{11}{2}(x - 3)$$

$$y - 7 = -\frac{11}{2}x + \frac{33}{2}$$

$$\underline{\qquad +\frac{11}{2}x \qquad\qquad +\frac{11}{2}x \qquad\qquad}$$

$$\frac{11}{2}x + y - 7 = \qquad \frac{33}{2}$$

$$\underline{\qquad\qquad +7 \qquad\qquad\qquad +7 \qquad}$$

$$\frac{11}{2}x + y = \qquad \frac{47}{2}$$

$$2\left(\frac{11}{2}x + y\right) = 2\left(\frac{47}{2}\right)$$

$$11x + 2y = 47 \qquad \text{(standard form)}$$

Linear Inequalities and Half-Planes

Each line plotted on a coordinate graph divides the graph (or plane) into two *half-planes*. This line is called the *boundary line* (or *bounding line*). The graph of a linear inequality is always a half-plane. Before graphing a linear inequality, you must first find or use the equation of the line to make a boundary line.

Open half-plane

If the inequality is a ">" or "<", then the graph will be an *open half-plane*. An open half-plane does not include the boundary line, so the boundary line is written as a *dashed line* on the graph.

Example 13: Graph the inequality $y < x - 3$.

First graph the line $y = x - 3$ to find the boundary line (use a dashed line, since the inequality is "<") as shown in Figure 9–12.

Figure 9–12 Graph of boundary line for $y < x - 3$.

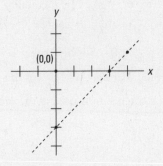

x	y
3	0
0	-3
4	1

Now shade the lower half-plane as shown in Figure 9–13, since $y < x - 3$.

Figure 9–13 Graph of inequality $y < x - 3$.

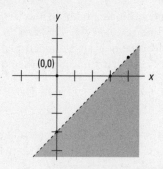

To check to see whether you've shaded the correct half-plane, plug in a pair of coordinates—the pair of (0, 0) is often a good choice. If the coordinates you selected make the *inequality a true statement* when plugged in, then you *should* shade the half-plane *containing* those coordinates. If the coordinates you selected *do not* make the inequality a true statement, then shade the half-plane *not containing* those coordinates.

Since the point (0, 0) *does not* make this inequality a true statement,

$$y < x - 3$$

$0 < 0 - 3$ is not true.

You should shade the side that *does not contain* the point (0, 0).

This checking method is often simply used as the method to decide which half-plane to shade.

Closed half-plane

If the inequality is a "≤" or "≥", then the graph will be a *closed half-plane*. A closed half-plane includes the boundary line and is graphed using a *solid line and shading*.

Example 14: Graph the inequality $2x - y \leq 0$.

First transform the inequality so that y is the left member.

Subtracting $2x$ from each side gives

$$-y \leq -2x$$

Now dividing each side by –1 (and changing the direction of the inequality) gives

$$y \geq 2x$$

Graph $y = 2x$ to find the boundary (use a solid line, because the inequality is "$\geq$") as shown in Figure 9–14.

Figure 9–14 Graph of the boundary line for $y \geq 2x$.

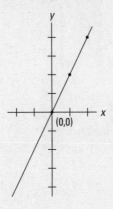

x	y
0	0
1	2
2	4

Since $y \geq 2x$, you should shade the upper half-plane. If in doubt, or to check, plug in a pair of coordinates. Try the pair (1, 1).

$$y \geq 2x$$

$$1 \geq 2(1)$$

$1 \geq 2$ is not true.

So you should shade the half-plane that *does not contain* (1, 1) as shown in Figure 9–15.

Figure 9–15 Graph of inequality $y \geq 2x$.

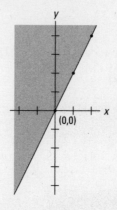

Chapter Check-Out

1. Is $x^2 - 8 = y$ linear or nonlinear?
2. Graph: $x + y = 8$

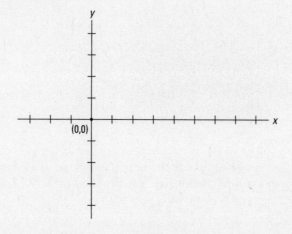

3. Find the equation of the line, in point-slope form, passing through the point (6, 1) with a slope of –2. Then convert it to standard form.
4. Find the equation of the line, in slope-intercept form, passing through the points (–3, 4) and (2, –3). Then convert it to standard form.

5. Graph: $y \le x - 2$

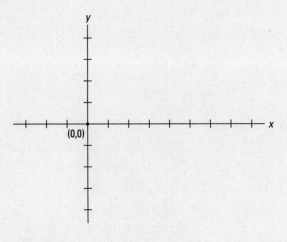

Answers: **1.** Nonlinear

2.

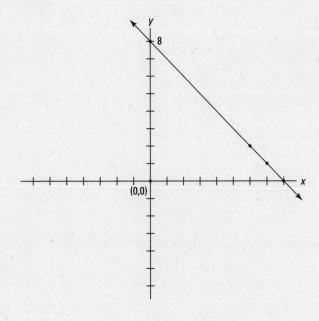

3. point-slope form: $y - 1 = -2(x - 6)$; standard form: $2x + y = 13$

4. slope-intercept form: $y = -\dfrac{7}{5}x - \dfrac{1}{5}$; standard form: $7x + 5y = -1$

5.

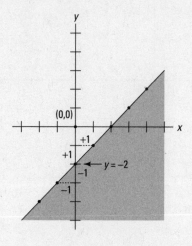

CHAPTER 10
FUNCTIONS AND VARIATIONS

Chapter Check-In

❑ Functions

❑ Finding domain and range

❑ Determining functions

❑ Finding values of functions

❑ Variations—direct and indirect

❑ Finding the constant of variation

Functions and variations deal with relationships between a set of values of one variable and a set of values of other variables. Specific definitions and examples given in this chapter will simplify what is often unnecessarily viewed as a difficult section.

Functions

Functions are very specific types of relations. Before defining a function, it is important to define a relation.

Relations

Any set of ordered pairs is called a **relation.** Figure 10–1 shows a set of ordered pairs.

$A = \{(-1, 1), (1, 3), (2, 2), (3, 4)\}$

Figure 10–1 A graph of the set of ordered pairs (–1, 1), (1, 3), (2, 2), (3, 4).

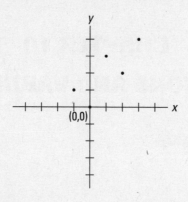

Domain and range

The set of all *x*'s is called the domain of the relation. The set of all *y*'s is called the range of the relation. The domain of set *A* in Figure 10-1 is {–1, 1, 2, 3}, while the range of set *A* is {1, 2, 3, 4}.

Example 1: Find the domain and range of the set of graphed points in Figure 10-2.

Figure 10–2 Plotted points.

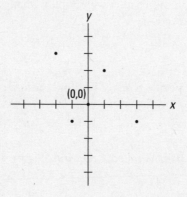

The domain is the set {–2, –1, 1, 3}. The range is the set {–1, 2, 3}.

Defining a function

The relation in Example 1 has pairs of coordinates with unique first terms. When the x value of each pair of coordinates is different, the relation is called a *function*. A function is a relation in which each member of the domain is paired with exactly one element of the range. *All functions are relations, but not all relations are functions.* A good example of a functional relation can be seen in the linear equation $y = x + 1$, graphed in Figure 10-3. The domain and range of this function are both the set of real numbers, and the relation is a function because for any value of x there is a unique value of y.

Figure 10–3 A graph of the linear equation $y = x + 1$.

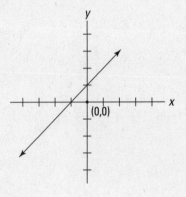

Graphs of functions

In each case in Figure 10-4 (a), (b), and (c), for any value of x, there is only one value for y. Contrast this with the graphs in Figure 10-5.

Figure 10-4 Graphs of functions.

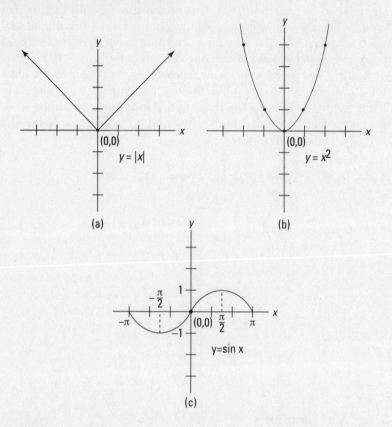

(a)

$y = |x|$

(b)

$y = x^2$

(c)

$y = \sin x$

Graphs of relationships that are not functions

In each of the relations in Figure 10-5 (a), (b), and (c), a single value of x is associated with two or more values of y. These relations are not functions.

Figure 10–5 Graphs of relations that are not functions.

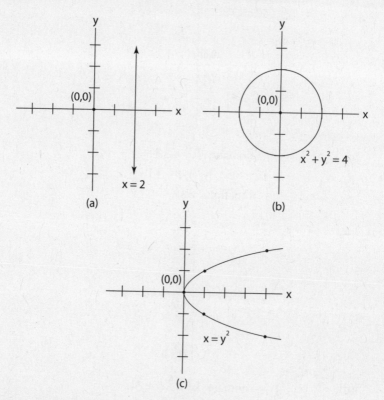

Determining domain, range, and if the relation is a function

Example 2:

(a) $B = \{(-2, 3),(-1, 4),(0, 5),(1, -3)\}$ domain: $\{-2, -1, 0, 1\}$

range: $\{-3, 3, 4, 5\}$

function: yes

(b)   domain: $\{-2, -1, 1, 2\}$

range: $\{-2, -1, 2\}$

function: yes

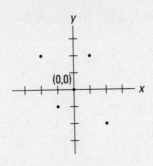

(c)   domain: $\{x\colon\ x \le 1\}$

range: $\{y\colon\ y \ge -3\ \}$

function: yes

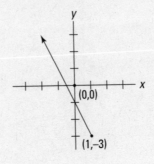

(d)   domain: $\{x\colon 0 \le x < 3\}$

range: $\{y\colon -2 < y < 2\}$

function: no

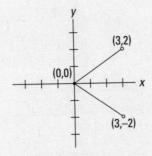

(e) $y = x^2$ domain: {all real numbers}

range: $\{y\colon y \ge 0\}$

function: yes

(f) $x = y^2$ domain: $\{x: x \geq 0\}$

range: {all real numbers}

function: no

Note that Examples 2(e) and (f) are illustrations of *inverse relations*: relations in which the domain and the range have been interchanged. Notice that while the relation in (e) is a function, the inverse relation in (f) is not.

Finding the values of functions

The *value of a function* is really the *value of the range* of the relation. Given the function

$$f = \{(1, -3),\ (2, 4),\ (-1, 5),\ (3, -2)\}$$

the value of the function at 1 is -3, at 2 is 4, and so forth. This is written $f(1) = -3$ and $f(2) = 4$ and is usually read, "f of 1 = -3 and f of 2 = 4." The lowercase letter f has been used here to indicate the concept of function, but any lowercase letter might have been used.

Example 3: Let $h = \{(3, 1), (2, 2), (1, -2), (-2, 3)\}$ Find each of the following.

(a) $h(3) =$ **(b)** $h(2) =$ **(c)** $h(1) =$ **(d)** $h(-2) =$

$h(3) = 1$ $h(2) = 2$ $h(1) = -2$ $h(-2) = 3$

Example 4: If $g(x) = 2x + 1$, find each of the following.

(a) $g(-1) =$ **(b)** $g(2) =$ **(c)** $g(a) =$

$g(x) = 2x + 1$ $g(x) = 2x + 1$ $g(x) = 2x + 1$

$g(-1) = 2(-1) + 1$ $g(2) = 2(2) + 1$ $g(a) = 2(a) + 1$

$g(-1) = -2 + 1$ $g(2) = 4 + 1$ $g(a) = 2a + 1$

$g(-1) = -1$ $g(2) = 5$

Example 5: If $f(x) = 3x^2 + x - 1$, find the range of f for the domain $\{-2, -1, 1\}$.

$f(x) = 3x^2 + x - 1$

$f(-2) = 3(-2)^2 + (-2) - 1$

$f(-2) = 3(4) - 2 - 1$

$f(-2) = 12 - 3$

$f(-2) = 9$

$$f(x) = 3x^2 + x - 1$$
$$f(-1) = 3(-1)^2 + (-1) - 1$$
$$f(-1) = 3(1) - 1 - 1$$
$$f(-1) = 3 - 2$$
$$f(-1) = 1$$
$$f(x) = 3x^2 + x - 1$$
$$f(1) = 3(1)^2 + (1) - 1$$
$$f(1) = 3(1) + 1 - 1$$
$$f(1) = 3$$

range: {9, 1, 3}

Variations

A **variation** is a relation between a set of values of one variable and a set of values of other variables.

Direct variation

In the equation $y = mx + b$, if m is a nonzero constant and $b = 0$, then you have the function $y = mx$ (often written $y = kx$), which is called a direct variation. That is, you can say that y varies directly as x or y is directly proportional to x. In this function, m (or k) is called the constant of proportionality or the constant of variation. The graph of every direct variation passes through the origin.

Example 6: Graph $y = 2x$.

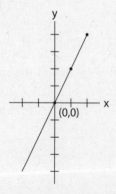

x	y
0	0
1	2
2	4

Example 7: If y varies directly as x, find the constant of variation when y is 2 and x is 4.

Because this is a direct variation,

$$y = kx \text{ (or } y = mx)$$

Now, replacing y with 2 and x with 4,

$$2 = k(4)$$

So
$$k = \frac{2}{4} \text{ or } k = \frac{1}{2}$$

The constant of variation is $\frac{1}{2}$.

Example 8: If y varies directly as x and the constant of variation is 2, find y when x is 6.

Since this is a direct variation, simply replace k with 2 and x with 6 in the following equation.

$$y = kx$$
$$y = 2(6)$$

So
$$y = 12$$

A direct variation can also be written as a proportion.

$$\frac{y_1}{x_1} = \frac{y_2}{x_2}$$

This proportion is read, "y_1 is to x_1 as y_2 is to x_2." x_1 and y_2 are called the **means,** and y_1 and x_2 are called the **extremes.** The product of the means is always equal to the product of the extremes. You can solve a proportion by simply multiplying the means and extremes and then solving as usual.

Example 9: r varies directly as p. If r is 3 when p is 7, find p when r is 9.

Method 1. Using proportions:

Set up the direct variation proportion

$$\frac{r_1}{p_1} = \frac{r_2}{p_2}$$

Now, substitute in the values.

$$\frac{3}{7} = \frac{9}{p}$$

Multiply the means and extremes (cross multiplying) give

$$3p = 63$$
$$p = 21$$

Method 2. Using $y = kx$:

Replace the y with p and the x with r.

$$p = kr$$

Use the first set of information and substitute 3 for r and 7 for p, then find k.

$$7 = k(3)$$
$$\frac{7}{3} = k \text{ or } k = \frac{7}{3}$$

Rewrite the direct variation equation as $p = \frac{7}{3}r$.

Now use the second set of information that says r is 9, substitute this into the preceding equation, and solve for p.

$$p = \frac{7}{3}r$$

$$p = \frac{7}{\cancel{3}}\left(\frac{\cancel{9}^{3}}{1}\right) = 21$$

Inverse variation (indirect variation)

A variation where $y = \dfrac{m}{x}$ or $y = \dfrac{k}{x}$ is called an *inverse variation* (or *indirect variation*). That is, *as x increases, y decreases*. And *as y increases, x decreases*. You may see the equation $xy = k$ representing an inverse variation, but this is simply a rearrangement of $y = \dfrac{k}{x}$.

This function is also referred to as an *inverse* or *indirect proportion*. Again, m (or k) is called the constant of variation.

Example 10: If y varies indirectly as x, find the constant of variation when y is 2 and x is 4.

Since this is an indirect or inverse variation,

$$y = \frac{k}{x}$$

Now, replacing y with 2 and x with 4,

$$2 = \frac{k}{4}$$

So $\qquad\qquad\qquad k = 2(4)$ or 8.

The constant of variation is 8.

Example 11: If y varies indirectly as x and the constant of variation is 2, find y when x is 6.

Since this is an indirect variation, simply replace k with 2 and x with 6 in the following equation.

$$y = \frac{k}{x}$$

$$y = \frac{2}{6}$$

So $\qquad\qquad\qquad y = \dfrac{1}{3}$

As in direct variation, inverse variation also can be written as a proportion.

$$\frac{y_1}{x_2} = \frac{y_2}{x_1}$$

Notice that in the inverse proportion, the x_1 and the x_2 switched their positions from the direct variation proportion.

Example 12: If y varies indirectly as x and $y = 4$ when $x = 9$, find x when $y = 3$.

Method 1. Using proportions:

Set up the indirect variation proportion.

$$\frac{y_1}{x_2} = \frac{y_2}{x_1}$$

Now, substitute in the values.

$$\frac{4}{x} = \frac{3}{9}$$

Multiply the means and extremes (cross-multiplying) gives

$$3x = 36$$

$$x = 12$$

Method 2. Using $y = \frac{k}{x}$:

Use the first set of information and substitute 4 for y and 9 for x, then find k.

$$4 = \frac{k}{9}$$

$$36 = k \quad \text{or} \quad k = 36$$

Rewrite the direct variation equation as $y = \frac{36}{x}$.

Now use the second set of information that says y is 3, substitute this into the preceding equation and solve for x.

$$3 = \frac{36}{x}$$

$$3x = 36$$

$$x = 12$$

Chapter Check-Out

1. Find out the range and domain of the set of ordered pairs: $\{(-2, 1), (-1, 3), (2, 4), (3, 5)\}$

2. Which of the following are graphs of functions?

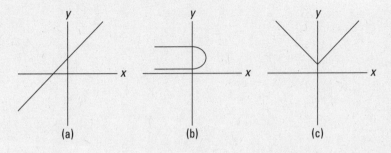

(a)   (b)   (c)

3. If $g(x) = 2x^2 + x + 1$, what is $g(-2)$?

4. If the domain of $x^2 + 2x + 2$ is $(1, 2, 3)$, then what is the range?

5. If y varies directly as x, find the constant of variation if y is 4 when x is 12.

6. r varies directly as p. If r is 6 and p is 11, find p when r is 18.

7. If y varies indirectly as x and the constant of variation is 3, find y when x is 7.

8. If y varies indirectly as x and y is 18 when x is 4, find y when x is 12.

Answers: 1. Range $\{1, 3, 4, 5\}$ Domain $\{-2, -1, 2, 3\}$ **2.** (a), (c) **3.** 7 **4.** $\{5, 10, 17\}$ **5.** $\frac{1}{3}$ **6.** 33 **7.** $\frac{3}{7}$ **8.** 6

CHAPTER 11

ROOTS AND RADICALS

Chapter Check-In

❏ Simplifying square roots

❏ Operations with square roots

❏ Using the conjugate

Note: This subject is introduced in the pre-algebra section (see Chapter 1).

The symbol $\sqrt{}$ is called a **radical sign** and is used to designate **square root**. To designate **cube root,** a small three is placed above the radical sign, $\sqrt[3]{}$. When two radical signs are next to each other, they automatically mean that the two are multiplied. The multiplication sign may be omitted. Note that the square root of a negative number is not possible within the real number system; a completely different system of **imaginary numbers** is used. The (so-called) imaginary numbers are multiples of the imaginary unit i.

$$\sqrt{-1} = i, \ \sqrt{-4} = 2i, \ \sqrt{-9} = 3i, \text{ and so on.}$$

Simplifying Square Roots

Example 1: Simplify.

(a) $\sqrt{9} = 3$ ⎫ Reminder: This notation is used in many

(b) $-\sqrt{9} = -3$ ⎭ texts and will be adhered to in this book.

(c) $\sqrt{18} = \sqrt{9 \cdot 2} = \sqrt{9} \cdot \sqrt{2} = 3\sqrt{2}$

(d) If each variable is nonnegative (not a negative number),

$\sqrt{x^2} = x$

If each variable could be positive or negative (deleting the restriction "If each variable is nonnegative"), then absolute value signs are placed around variables to odd powers.

$$\sqrt{x^2} = |x|$$

(e) If each variable is nonnegative,

$$\sqrt{x^4} = x^2$$

(f) If each variable is nonnegative,

$$\sqrt{x^6 y^8} = \sqrt{x^6}\sqrt{y^8} = x^3 y^4$$

If each variable could be positive or negative, then you would write

$$|x^3|y^4$$

(g) If each variable is nonnegative,

$$\sqrt{25a^4 b^6} = \sqrt{25}\sqrt{a^4}\sqrt{b^6} = 5a^2 b^3$$

If each variable could be positive or negative, you would write

$$5a^2 |b^3|$$

(h) If each variable is nonnegative,

$$\sqrt{x^7} = \sqrt{x^6 (x)} = \sqrt{x^6}\sqrt{x} = x^3 \sqrt{x}$$

If each variable could be positive or negative, you would write

$$|x^3|\sqrt{x}$$

(i) If each variable is nonnegative,

$$\sqrt{x^9 y^8} = \sqrt{x^9}\sqrt{y^8} = \sqrt{x^8 (x)}\sqrt{y^8} = \sqrt{x^8}\sqrt{x}\sqrt{y^8} = x^4 \sqrt{x} \cdot y^4 = x^4 y^4 \sqrt{x}$$

(j) If each variable is nonnegative,

$$\sqrt{16x^5} = \sqrt{16}\sqrt{x^5} = \sqrt{16}\sqrt{x^4 (x)} = \sqrt{16}\sqrt{x^4}\sqrt{x} = 4x^2 \sqrt{x}$$

Operations with Square Roots

You can perform a number of different operations with square roots. Some of these operations involve a single radical sign, while others can involve many radical signs. The rules governing these operations should be carefully reviewed.

Under a single radical sign

You may perform operations *under a single radical sign*.

Example 2: Perform the operation indicated.

(a) $\sqrt{(5)(20)} = \sqrt{100} = 10$

(b) $\sqrt{30+6} = \sqrt{36} = 6$

(c) $\sqrt{\dfrac{32}{2}} = \sqrt{16} = 4 \left(\textbf{Note}: \sqrt{\dfrac{32}{2}} = \dfrac{\sqrt{32}}{\sqrt{2}} \right)$

(d) $\sqrt{30-5} = \sqrt{25} = 5$

(e) $\sqrt{2+5} = \sqrt{7}$

When radical values are alike

You can *add or subtract square roots themselves only if the values under the radical sign are equal.* Then simply add or subtract the coefficients (numbers in front of the radical sign) and keep the original number in the radical sign.

Example 3: Perform the operation indicated.

(a) $2\sqrt{3} + 3\sqrt{3} = (2+3)\sqrt{3} = 5\sqrt{3}$

(b) $4\sqrt{6} - 2\sqrt{6} = (4-2)\sqrt{6} = 2\sqrt{6}$

(c) $5\sqrt{2} + \sqrt{2} = 5\sqrt{2} + 1\sqrt{2} = (5+1)\sqrt{2} = 6\sqrt{2}$

Note that the coefficient 1 is understood in $\sqrt{2}$.

When radical values are different

You may not add or subtract different square roots.

Example 4:

(a) $\sqrt{28} - \sqrt{3} \neq \sqrt{25}$

(b) $\sqrt{16} + \sqrt{9} \neq \sqrt{25}$

$\qquad 4 + 3 \neq 5$

Addition and subtraction of square roots after simplifying

Sometimes, after simplifying the square root(s), addition or subtraction becomes possible. Always simplify if possible.

Example 5: Simplify and add.

(a) $\sqrt{50} + 3\sqrt{2} =$

These cannot be added until $\sqrt{50}$ is simplified.

$$\sqrt{50} = \sqrt{25 \cdot 2} = \sqrt{25} \cdot \sqrt{2} = 5\sqrt{2}$$

Now, because both are alike under the radical sign,

$$5\sqrt{2} + 3\sqrt{2} = (5+3)\sqrt{2} = 8\sqrt{2}$$

(b) $\sqrt{300} + \sqrt{12} =$

Try to simplify each one.

$$\sqrt{300} = \sqrt{100 \cdot 3} = \sqrt{100} \cdot \sqrt{3} = 10\sqrt{3}$$

$$\sqrt{12} = \sqrt{4 \cdot 3} = \sqrt{4} \cdot \sqrt{3} = 2\sqrt{3}$$

Now, because both are alike under the radical sign,

$$10\sqrt{3} + 2\sqrt{3} = (10+2)\sqrt{3} = 12\sqrt{3}$$

Products of nonnegative roots

Remember that in multiplication of roots, the multiplication sign may be omitted. Always simplify the answer when possible.

Example 6: Multiply.

(a) $\sqrt{2} \cdot \sqrt{8} = \sqrt{16} = 4$

(b) If each variable is nonnegative,

$$\sqrt{x^3} \cdot \sqrt{x^5} = \sqrt{x^8} = x^4$$

(c) If each variable is nonnegative,

$$\sqrt{ab}\cdot\sqrt{ab^3c}=\sqrt{a^2b^4c}=\sqrt{a^2}\sqrt{b^4}\sqrt{c}=ab^2\sqrt{c}$$

(d) If each variable is nonnegative,

$$\sqrt{3x}\cdot\sqrt{6xy^2}\cdot\sqrt{2xy}=\sqrt{36x^3y^3}=\sqrt{36}\sqrt{x^3}\sqrt{y^3}=$$

$$\sqrt{36}\sqrt{x^2(x)}\sqrt{y^2(y)}=\sqrt{36}\sqrt{x^2}\sqrt{x}\sqrt{y^2}\sqrt{y}=6xy\sqrt{xy}$$

(e) $2\sqrt{5}\cdot7\sqrt{3}=(2\cdot7)\sqrt{5\cdot3}=14\sqrt{15}$

Quotients of nonnegative roots

For all positive numbers,

$$\frac{\sqrt{x}}{\sqrt{y}}=\sqrt{\frac{x}{y}}$$

In the following examples, all variables are assumed to be positive.

Example 7: Divide. Leave all fractions with rational denominators.

(a) $\dfrac{\sqrt{10}}{\sqrt{2}}=\sqrt{\dfrac{10}{2}}=\sqrt{5}$

(b) $\dfrac{\sqrt{24}}{\sqrt{3}}=\sqrt{\dfrac{24}{3}}=\sqrt{8}=2\sqrt{2}$

(c) $\dfrac{\sqrt{28x^6}}{\sqrt{7x^2}}=\sqrt{\dfrac{28x^6}{7x^2}}=\sqrt{4x^4}=2x^2$

(d) $\dfrac{\sqrt{15}}{\sqrt{6}}=\sqrt{\dfrac{15}{6}}=\sqrt{\dfrac{5}{2}}$ or $\dfrac{\sqrt{5}}{\sqrt{2}}$

Note that the denominator of this fraction in part (d) is irrational. In order to rationalize the denominator of this fraction, multiply it by 1 in the form of

$$1=\frac{\sqrt{2}}{\sqrt{2}}$$

$$\frac{\sqrt{5}}{\sqrt{2}}\cdot1=\frac{\sqrt{5}}{\sqrt{2}}\cdot\frac{\sqrt{2}}{\sqrt{2}}=\frac{\sqrt{10}}{\sqrt{4}}=\frac{\sqrt{10}}{2}$$

Example 8: Divide. Leave all fractions with rational denominators.

(a) $\dfrac{5\sqrt{7}}{\sqrt{12}}$

First simplify $\sqrt{12}$:

$$\frac{5\sqrt{7}}{\sqrt{12}} = \frac{5\sqrt{7}}{2\sqrt{3}} \cdot 1 = \frac{5\sqrt{7}}{2\sqrt{3}} \cdot \frac{\sqrt{3}}{\sqrt{3}} = \frac{5\sqrt{21}}{2\sqrt{9}} = \frac{5\sqrt{21}}{2 \cdot 3} = \frac{5\sqrt{21}}{6}$$

or

$$\frac{5\sqrt{7}}{\sqrt{12}} \cdot \frac{\sqrt{12}}{\sqrt{12}} = \frac{5\sqrt{7} \cdot \sqrt{12}}{12} = \frac{5\sqrt{84}}{12} = \frac{5\sqrt{4 \cdot 21}}{12} =$$

$$\frac{10\sqrt{21}}{12} = \frac{5\sqrt{21}}{6}$$

(b) $\dfrac{9\sqrt{2x}}{\sqrt{24x^3}} = 9\sqrt{\dfrac{2x}{24x^3}} = \dfrac{9}{\sqrt{12x^2}} = \dfrac{9}{2x\sqrt{3}} \cdot 1 =$

$$\frac{9}{2x\sqrt{3}} \cdot \frac{\sqrt{3}}{\sqrt{3}} = \frac{9\sqrt{3}}{2x \cdot 3} = \frac{9\sqrt{3}}{6x} = \frac{3\sqrt{3}}{2x}$$

(c) $\dfrac{3}{2+\sqrt{3}} \cdot 1 = \dfrac{3}{2+\sqrt{3}} \cdot \dfrac{\left(2-\sqrt{3}\right)}{\left(2-\sqrt{3}\right)} =$

$$\frac{3\left(2-\sqrt{3}\right)}{4-3} = \frac{6-3\sqrt{3}}{1} = 6-3\sqrt{3}$$

Note: In order to leave a rational term in the denominator, it is necessary to multiply both the numerator and denominator by the *conjugate* of the denominator. The conjugate of a binomial contains the same terms but the opposite sign. Thus, $(x + y)$ and $(x - y)$ are conjugates.

Example 9: Divide. Leave the fraction with a rational denominator.

$$\frac{1+\sqrt{5}}{2-\sqrt{5}} \cdot 1 = \frac{\left(1+\sqrt{5}\right)}{\left(2-\sqrt{5}\right)} \cdot \frac{\left(2+\sqrt{5}\right)}{\left(2+\sqrt{5}\right)} = \frac{2+\sqrt{5}+2\sqrt{5}+\sqrt{25}}{4+2\sqrt{5}-2\sqrt{5}-\sqrt{25}} =$$

$$\frac{2+3\sqrt{5}+5}{4-5} = \frac{7+3\sqrt{5}}{-1} = -7-3\sqrt{5}$$

Chapter Check-Out

1. Simplify: $\sqrt{50}$

2. If each variable is nonnegative $\sqrt{16a^6b^8} =$

3. If each variable is positive or negative, then $\sqrt{x^7} =$

4. $\sqrt{(4)(36)} =$

5. $\sqrt{60} + 2\sqrt{15} =$

6. $\sqrt{6} \times \sqrt{10} =$

7. If x is nonnegative, then $\sqrt{x^3 y} \cdot \sqrt{x^2 y} =$

For questions 8–10, express answers in simplest form with rationalized denominators.

8. $\dfrac{\sqrt{28}}{\sqrt{7}} =$

9. $\dfrac{3\sqrt{5}}{\sqrt{2}} =$

10. $\dfrac{2}{3+\sqrt{2}} =$

Answers: 1. $5\sqrt{2}$ **2.** $4a^3b^4$ **3.** $\left|x^3\right|\sqrt{x}$ **4.** 12 **5.** $4\sqrt{15}$ **6.** $2\sqrt{15}$ **7.** $x^2 y\sqrt{x}$ **8.** 2 **9.** $\dfrac{3\sqrt{10}}{2}$ **10.** $\dfrac{6-2\sqrt{2}}{7}$

CHAPTER 12

QUADRATIC EQUATIONS

Chapter Check-In

❑ Solving quadratic equations

❑ The factoring method

❑ The quadratic formula

❑ Completing the square

A quadratic equation is an equation that could be written as

$$ax^2 + bx + c = 0$$

when $a \neq 0$.

Solving Quadratic Equations

There are three basic methods for solving quadratic equations: factoring, using the quadratic formula, and completing the square. For a review of factoring, refer to Chapter 6.

Factoring

To solve a quadratic equation by factoring,

1. Put all terms on one side of the equal sign, leaving zero on the other side.
2. Factor.
3. Set each factor equal to zero.
4. Solve each of these equations.
5. Check by inserting your answer in the original equation.

Example 1: Solve $x^2 - 6x = 16$.

Following the steps,

$$x^2 - 6x = 16 \text{ becomes } x^2 - 6x - 16 = 0$$

Factor.

$$(x - 8)(x + 2) = 0$$

Setting each factor to zero,

$$x - 8 = 0 \qquad \text{or} \qquad x + 2 = 0$$
$$x = 8 \qquad\qquad\qquad x = -2$$

Then to check,

$$8^2 - 6(8) = 16 \qquad \text{or} \qquad (-2)^2 - 6(-2) = 16$$
$$64 - 48 = 16 \qquad\qquad\qquad 4 + 12 = 16$$
$$16 = 16 \qquad\qquad\qquad 16 = 16$$

Both values, 8 and –2, are solutions to the original equation.

Example 2: Solve $y^2 = -6y - 5$.

Setting all terms equal to zero,

$$y^2 + 6y + 5 = 0$$

Factor.

$$(y + 5)(y + 1) = 0$$

Setting each factor to 0,

$$y + 5 = 0 \qquad \text{or} \qquad y + 1 = 0$$
$$y = -5 \qquad\qquad\qquad y = -1$$

To check, $y^2 = -6y - 5$

$$(-5)^2 = -6(-5) - 5 \qquad \text{or} \qquad (-1)^2 = -6(-1) - 5$$
$$25 = 30 - 5 \qquad\qquad\qquad 1 = 6 - 5$$
$$25 = 25 \qquad\qquad\qquad 1 = 1$$

A quadratic with a term missing is called an **incomplete quadratic** (as long as the ax^2 term isn't missing).

Example 3: Solve $x^2 - 16 = 0$.

Factor.

$$(x + 4)(x - 4) = 0$$

$$x + 4 = 0 \qquad \text{or} \qquad x - 4 = 0$$

$$x = -4 \qquad\qquad\qquad x = 4$$

To check, $x^2 - 16 = 0$

$$(-4)^2 - 16 = 0 \qquad \text{or} \qquad (4)^2 - 16 = 0$$

$$16 - 16 = 0 \qquad\qquad\qquad 16 - 16 = 0$$

$$0 = 0 \qquad\qquad\qquad\qquad 0 = 0$$

Example 4: Solve $x^2 + 6x = 0$.

Factor.

$$x(x + 6) = 0$$

$$x = 0 \qquad \text{or} \qquad x + 6 = 0$$

$$x = 0 \qquad\qquad\qquad x = -6$$

To check, $x^2 + 6x = 0$

$$(0)^2 + 6(0) = 0 \qquad \text{or} \qquad (-6)^2 + 6(-6) = 0$$

$$0 + 0 = 0 \qquad\qquad\qquad 36 + (-36) = 0$$

$$0 = 0 \qquad\qquad\qquad\qquad 0 = 0$$

Example 5: Solve $2x^2 + 2x - 1 = x^2 + 6x - 5$.

First, simplify by putting all terms on one side and combining like terms.

$$2x^2 + 2x - 1 = x^2 + 6x - 5$$
$$\underline{-x^2 - 6x + 5 \quad -x^2 - 6x + 5}$$
$$x^2 - 4x + 4 = 0$$

Now, factor.

$$(x - 2)(x - 2) = 0$$

$$x - 2 = 0$$

$$x = 2$$

To check, $2x^2 + 2x - 1 = x^2 + 6x - 5$

$$2(2)^2 + 2(2) - 1 = (2)^2 + 6(2) - 5$$
$$8 + 4 - 1 = 4 + 12 - 5$$
$$11 = 11$$

The quadratic formula

Many quadratic equations cannot be solved by factoring. This is generally true when the roots, or answers, are not rational numbers. A second method of solving quadratic equations involves the use of the following formula:

$$x = \frac{-b \pm \sqrt{b^2 - 4ac}}{2a}$$

a, b, and *c* are taken from the quadratic equation written in its general form of

$$ax^2 + bx + c = 0$$

where *a* is the numeral that goes in front of x^2, *b* is the numeral that goes in front of *x*, and *c* is the numeral with no variable next to it (a.k.a., "the constant").

When using the quadratic formula, you should be aware of three possibilities. These three possibilities are distinguished by a part of the formula called the discriminant. The *discriminant* is the value under the radical sign, $b^2 - 4ac$. A quadratic equation with real numbers as coefficients can have the following:

1. Two different real roots if the discriminant $b^2 - 4ac$ is a positive number.
2. One real root if the discriminant $b^2 - 4ac$ is equal to 0.
3. No real root if the discriminant $b^2 - 4ac$ is a negative number.

Example 6: Solve for x: $x^2 - 5x = -6$.

Setting all terms equal to 0,

$$x^2 - 5x + 6 = 0$$

Then substitute 1 (which is understood to be in front of the x^2), –5, and 6 for a, b, and c, respectively, in the quadratic formula and simplify.

$$x = \frac{-b \pm \sqrt{b^2 - 4ac}}{2a}$$

$$x = \frac{-(-5) \pm \sqrt{(-5)^2 - 4(1)(6)}}{2(1)}$$

$$x = \frac{5 \pm \sqrt{25 - 24}}{2}$$

$$x = \frac{5 \pm \sqrt{1}}{2}$$

$$x = \frac{5 \pm 1}{2}$$

$$x = \frac{5 + 1}{2} \text{ or } x = \frac{5 - 1}{2}$$

$$x = \frac{6}{2} \text{ or } x = \frac{4}{2}$$

$$x = 3 \text{ or } x = 2$$

Because the discriminant $b^2 - 4ac$ is positive, you get two different real roots.

Example 6 produces rational roots. In Example 7, the quadratic formula is used to solve an equation whose roots are not rational.

Example 7: Solve for y: $y^2 = -2y + 2$.

Setting all terms equal to 0,

$$y^2 + 2y - 2 = 0$$

Then substitute 1, 2, and –2 for a, b, and c, respectively, in the quadratic formula and simplify.

$$y = \frac{-b \pm \sqrt{b^2 - 4ac}}{2a}$$

$$y = \frac{-(2) \pm \sqrt{(2)^2 - 4(1)(-2)}}{2(1)}$$

$$y = \frac{-2 \pm \sqrt{4 + 8}}{2}$$

$$y = \frac{-2 \pm \sqrt{12}}{2}$$

$$y = \frac{-2 \pm \sqrt{4}\sqrt{3}}{2}$$

$$y = \frac{-2 \pm 2\sqrt{3}}{2}$$

$$y = \frac{\overset{1}{2}\left(-1 \pm \sqrt{3}\right)}{\underset{1}{2}}$$

$$y = -1 + \sqrt{3} \text{ or } y = -1 - \sqrt{3}$$

Note that the two roots are irrational.

Example 8: Solve for x: $x^2 + 2x + 1 = 0$.

Substituting in the quadratic formula,

$$x = \frac{-b \pm \sqrt{b^2 - 4ac}}{2a}$$

$$x = \frac{-(2) \pm \sqrt{(2)^2 - 4(1)(1)}}{2(1)}$$

$$x = \frac{-2 \pm \sqrt{4 - 4}}{2}$$

$$x = \frac{-2 \pm \sqrt{0}}{2}$$

$$x = \frac{-2}{2} = -1$$

Since the discriminant $b^2 - 4ac$ is 0, the equation has one root.

The quadratic formula can also be used to solve quadratic equations whose roots are imaginary numbers, that is, they have no solution in the real number system.

Example 9: Solve for x: $x(x + 2) + 2 = 0$, or $x^2 + 2x + 2 = 0$.

Substituting in the quadratic formula,

$$x = \frac{-b \pm \sqrt{b^2 - 4ac}}{2a}$$

$$x = \frac{-(2) \pm \sqrt{(2)^2 - 4(1)(2)}}{2(1)}$$

$$x = \frac{-2 \pm \sqrt{4 - 8}}{2}$$

$$x = \frac{-2 \pm \sqrt{-4}}{2}$$

Since the discriminant $b^2 - 4ac$ is negative, this equation has no solution in the real number system.

But if you were to express the solution using imaginary numbers, as discussed at the beginning of Chapter 11, the solutions would be $x = \frac{-2 \pm \sqrt{-4}}{2} = \frac{-2 \pm 2i}{2} = -1 \pm i$ or $-1 + i$, $-1 - i$.

Completing the square

A third method of solving quadratic equations that works with both real and imaginary roots is called completing the square.

1. Put the equation into the form $ax^2 + bx = -c$.
2. Make sure that $a = 1$ (if $a \neq 1$, multiply through the equation by $\frac{1}{a}$ before proceeding).
3. Using the value of b from this new equation, add $\left(\frac{b}{2}\right)^2$ to both sides of the equation to form a perfect square on the left side of the equation.
4. Find the square root of both sides of the equation.
5. Solve the resulting equation.

Example 10: Solve for x: $x^2 - 6x + 5 = 0$.

Arrange in the form of

$$ax^2 + bx = -c$$

$$x^2 - 6x = -5$$

Because $a = 1$, add $\left(\dfrac{-6}{2}\right)^2$, or 9, to both sides to complete the square.

$$x^2 - 6x + 9 = -5 + 9$$

$$x^2 - 6x + 9 = 4$$

$$(x - 3)^2 = 4$$

Take the square root of both sides.

$$x - 3 = \pm 2$$

Solve.

$$x - 3 = \pm 2$$

$$\begin{array}{ccc} x - 3 = +2 & \text{or} & x - 3 = -2 \\ \underline{+3 \quad +3} & & \underline{+3 \quad +3} \\ x = 5 & \text{or} & x = 1 \end{array}$$

Example 11: Solve for y: $y^2 + 2y - 4 = 0$.

Arrange in the form of

$$ay^2 + by = -c$$

$$y^2 + 2y = 4$$

Because $a = 1$, add $\left(\dfrac{2}{2}\right)^2$, or 1, to both sides to complete the square.

$$y^2 + 2y + 1 = 4 + 1$$

$$y^2 + 2y + 1 = 5$$

or

$$(y + 1)^2 = 5$$

Take the square root of both sides.

$$y + 1 = \pm\sqrt{5}$$

Solve.

$$y + 1 = \pm\sqrt{5}$$

$$\underline{-1 \quad -1}$$

$$y = -1 \pm \sqrt{5}$$

$$y = -1 + \sqrt{5} \text{ or } y = -1 - \sqrt{5}$$

Example 12: Solve for x: $2x^2 + 3x + 2 = 0$.

Arrange in the form of

$$ax^2 + bx = -c$$

$$2x^2 + 3x = -2$$

Because $a \neq 1$, multiply through the equation by $\frac{1}{2}$.

$$x^2 + \frac{3}{2}x = -1$$

Add $\left[\left(\frac{1}{2}\right)\left(\frac{3}{2}\right)\right]^2$, or $\frac{9}{16}$ to both sides.

$$x^2 + \frac{3}{2}x + \frac{9}{16} = -1 + \frac{9}{16}$$

$$x^2 + \frac{3}{2}x + \frac{9}{16} = -\frac{7}{16}$$

$$\left(x + \frac{3}{4}\right)^2 = -\frac{7}{16}$$

Take the square root of both sides.

$$x + \frac{3}{4} = \pm\sqrt{\frac{-7}{16}}$$

$$x + \frac{3}{4} = \pm\frac{\sqrt{-7}}{\sqrt{16}}$$

$$x + \frac{3}{4} = \pm\frac{\sqrt{-7}}{4}$$

$$x = -\frac{3}{4} \pm \frac{\sqrt{-7}}{4}$$

There is no solution in the real number system. It may interest you to know that the completing the square process for solving quadratic equations was used on the equation $ax^2 + bx + c = 0$ to derive the quadratic formula.

Chapter Check-Out

1. Solve for x: $x^2 + 3x + 2 = 0$

2. Solve for x: $x^2 - 5x = 6$

3. Solve for x: $x^2 = 5x - 4$

4. Solve for x: $x^2 - 49 = 0$

5. Solve for x: $x^2 - 6x = 0$

6. Solve for x: $3x^2 + 21x = 2x^2 - 3x + 81$

7. Solve for x by using the quadratic formula: $x^2 + 5x + 2 = 0$

8. Solve for x by completing the square: $x^2 + 4x + 2 = 0$

Answers: 1. $-1, -2$ **2.** $6, -1$ **3.** $4, 1$ **4.** $7, -7$ **5.** $0, 6$ **6.** $3, -27$

7. $\dfrac{-5 \pm \sqrt{17}}{2}$ or $\dfrac{-5 + \sqrt{17}}{2}, \dfrac{-5 - \sqrt{17}}{2}$ **8.** $-2 \pm \sqrt{2}$ or $-2 + \sqrt{2}, -2 - \sqrt{2}$

CHAPTER 13

WORD PROBLEMS

Chapter Check-In

❑ Solving techniques

❑ Key words and phrases

❑ Simple and compound interest

❑ Ratio, proportion, and percent

❑ Number, age, and coin problems

❑ Motion, mixture, and work problems

Word problems are often the nemesis of even the best math student. For many, the difficulty is not the computation. The problems stem from what is given and what is being asked.

Solving Technique

There are many types of word problems involving arithmetic, algebra, geometry, and combinations of them with various twists. It is most important to have a systematic technique for solving word problems. Here is such a technique.

1. **First, identify what is being asked.** What are you ultimately trying to find? How far a car has traveled? How fast a plane flies? How many items can be purchased? Whatever it is, find it and then *circle it.* This helps ensure that you are solving for what is being asked.

2. **Next, underline and pull out information you are given in the problem.** Draw a picture if you can. This helps you know what you have and will point you to a relationship or equation. Note any key words in the problem (see "Key Words and Phrases" later in this chapter).

3. **If you can, set up an equation or some straightforward system with the given information.**

4. **Is all the given information necessary to solve the problem?** Occasionally, you may be given more than enough information to solve a problem. *Choose what you need* and don't spend needless energy on irrelevant information.

5. **Carefully solve the equation or work the necessary computation.** Be sure you are working in the same units (for example, you may have to change feet into inches, pounds into ounces, and so forth, in order to keep everything consistent).

6. **Did you answer the question?** One of the most common errors in answering word problems is the failure to answer what was actually being asked.

7. **And finally, is your answer reasonable?** Check to make sure that an error in computation or a mistake in setting up your equation did not give you a ridiculous answer.

Key Words and Phrases

In working with word problems, there are some words or phrases that give clues as to how the problem should be solved. The most common words or phrases are as follows.

- Add

    Sum—as in *the sum of 2, 3, and 6 . . .*

    Total—as in *the total of the first six payments . . .*

    Addition—as in *a recipe calls for the addition of five pints . . .*

    Plus—as in *three liters plus two liters . . .*

    Increase—as in *her pay was increased by $15 . . .*

    More than—as in *this week the enrollment was eight more than last week . . .*

    Added to—as in *if you added $3 to the cost . . .*

- **Subtract**

    Difference—as in *what is the difference between . . .*

    Fewer—as in *there were fifteen fewer men than women . . .*

Remainder—as in *how many are left or what quantity remains . . .*

Less than—as in *a number is five less than another number . . .*

Reduced—as in *the budget was reduced by $5,000 . . .*

Decreased—as in *if he decreased the speed of his car by ten miles per hour . . .*

Minus—as in *some number minus 9 is . . .*

■ **Multiply**

Product—as in *the product of 8 and 5 is . . .*

Of—as in *one-half of the group . . .*

Times—as in *five times as many girls as boys . . .*

At—as in *the cost of ten yards of material at 70¢ a yard is . . .*

Total—as in *if you spend $15 a week on gas, what is the total for a three-week period . . .*

Twice—as in *twice the value of some number . . .*

■ **Divide**

Quotient—as in *the final quotient is . . .*

Divided by—as in *some number divided by 12 is . . .*

Divided into—as in *the group was divided into . . .*

Ratio—as in *what is the ratio of . . .*

Half—as in *half the profits are . . .* (dividing by 2)

As you work a variety of word problem types, you will discover more "clue" words.

A final reminder: Be sensitive to what each of these questions is asking. What time? How many? How much? How far? How old? What length? What is the ratio?

Simple Interest

Example 1: How much simple interest will an account earn in five years if $500 is invested at 8% interest per year?

First, circle what you must find—*interest*. Now use the equation

Interest = principal times rate times time

$$I = prt$$

Simply plug into the equation.

$$I = \$500(.08)5$$

$$I = \$200$$

Note that both rate and time are in yearly terms (annual rate; years).

Compound Interest

Example 2: What will be the final total amount of money after three years on an original investment of $1,000 if a 12% annual interest rate is compounded yearly?

First, circle what you must find—*final total amount of money*. Note also that interest will be *compounded each year*. Therefore, the solution has three parts, one for each year.

Total for first year: *Interest = principal* times *rate* times *time*

$$I = prt$$

$$I = \$1,000 \times .12 \times 1$$

$$I = \$120$$

Thus, the total after one year is $1,000 + $120 = $1,120.

Total for second year: $I = prt$

$$I = \$1,120(.12)1$$

$$I = \$134.40$$

Note that the principal at the beginning of the second year was $1,120. Thus, the total after two years is $1,120 + $134.40 = $1,254.40.

Total for third year: $I = prt$

$$I = \$1,254.40(.12)1$$

$$I = \$150.53$$

Note that the principal at the beginning of the third year was $1,254.40. Thus, the total after three years is $1,254.40 + $150.53 = $1,404.93.

Ratio and Proportion

Example 3: If Arnold can type 600 pages of manuscript in 21 days, how many days will it take him to type 230 pages if he works at the same rate?

First, circle what you're asked to find—*how many days.* One simple way to work this problem is to set up a "framework" (proportion) using the categories given in the equation. Here the categories are *pages* and *days.*

Therefore, a framework may be

$$\frac{\text{pages}}{\text{days}} = \frac{\text{pages}}{\text{days}}$$

Note that you also may have used

$$\frac{\text{days}}{\text{pages}} = \frac{\text{days}}{\text{pages}}$$

The answer will be the same. Now simply plug into the equation for each instance.

$$\frac{600}{21} = \frac{230}{x}$$

Cross multiplying

$$600x = 21(230)$$
$$600x = 4,830$$
$$\frac{600x}{600} = \frac{4,830}{600}$$
$$x = 8\frac{1}{20} \text{ or } 8.05$$

Therefore, it will take $8\frac{1}{20}$ or 8.05 days to type 230 pages. (You may have simplified the original proportion before solving.)

Percent

Example 4: Thirty students are awarded doctoral degrees at the graduate school, and this number comprises 40% of the total graduate student body. How many graduate students were enrolled?

First, circle what you must find in the problem—*how many graduate students*. Now, in order to plug into the percentage equation

$$\frac{\text{is}}{\text{of}} = \%$$

try rephrasing the question into a simple sentence. For example, in this case,

30 is 40% of what total?

Notice that the 30 sits next to the word is; therefore, 30 is the "is" number. 40 is the percent. Notice that *what total* sits next to the word *of*. Therefore, plugging into the equation,

$$\frac{\text{is}}{\text{of}} = \%$$

$$\frac{30}{x} = \frac{40}{100}$$

Cross multiplying,

$$40x = 3{,}000$$

$$\frac{40x}{40} = \frac{3{,}000}{40}$$

$$x = 75$$

Therefore, the total graduate enrollment was 75 students.

Percent Change

To find the *percent change* (increase or decrease), use the formula given in Example 5.

Example 5: Last year, Harold earned $250 a month at his after-school job. This year, his after-school earnings have increased to $300 per month. What is the percent increase in his monthly after-school earnings?

First, circle what you're looking for—*percent increase*. Percent change (percent increase, percentage rise, % difference, percent decrease, and so forth) is always found by using the equation

$$\text{percent change} = \frac{\text{change}}{\text{starting point}}$$

Therefore,

$$\begin{aligned}
\text{percent change} &= \frac{\$300 - \$250}{\$250} \\
&= \frac{\$50}{\$250} \\
&= \frac{1}{5} = 0.20 = 20\%
\end{aligned}$$

The percent increase in Harold's after-school salary is 20%.

Number Problems

Example 6: When 6 times a number is increased by 4, the result is 40. Find the number.

First, circle what you must find—*the number*. Letting x stand for the number gives the equation

$$6x + 4 = 40$$

Subtracting 4 from each side gives

$$6x = 36$$

Dividing by 6 gives

$$x = 6$$

So the number is 6.

Example 7: One number exceeds another number by 5. If the sum of the two numbers is 39, find the smaller number.

First, circle what you are looking for—*the smaller number*. Now, let the smaller number equal x. Therefore, the larger number equals $x + 5$. Now, use the problem to set up an equation.

$$\underbrace{\text{If the sum of the two numbers}}_{x+(x+5)} \underbrace{\text{is}}_{=} \underbrace{39\dots}_{39}$$

$$2x + 5 = 39$$
$$2x + 5 - 5 = 39 - 5$$
$$2x = 34$$
$$\frac{2x}{2} = \frac{34}{2}$$
$$x = 17$$

Therefore, the smaller number is 17.

Example 8: If one number is three times as large as another number and the smaller number is increased by 19, the result is 6 less than twice the larger number. What is the larger number?

First, circle what you must find—*the larger number*. Let the smaller number equal x. Therefore, the larger number will be $3x$. Now, using the problem, set up an equation.

$$\underbrace{\text{The smaller number}}_{x} \underbrace{\text{increased by}}_{+} \underbrace{19}_{19} \underbrace{\text{is}}_{=} \underbrace{\text{6 less than twice the larger number.}}_{2(3x)-6}$$

$$x + 19 = 6x - 6$$
$$-x + x + 19 = -x + 6x - 6$$
$$19 = 5x - 6$$
$$19 + 6 = 5x - 6 + 6$$
$$25 = 5x$$
$$5 = x$$

Therefore, the larger number, $3x$, is $3(5)$, or 15.

Example 9: The sum of three consecutive integers is 306. What is the largest integer?

First, circle what you must find—*the largest integer.* Let the smallest integer equal x; let $x + 1$ equal the next integer; let the largest integer equal $x + 2$. Now, use the problem to set up an equation.

$$\underbrace{\text{The sum of the three consecutive integers}}_{x+(x+1)+(x+2)} \overset{\text{is}}{=} \underbrace{306.}_{306}$$

$$3x + 3 = 306$$
$$3x + 3 - 3 = 306 - 3$$
$$3x = 303$$
$$\frac{3x}{3} = \frac{303}{3}$$
$$x = 101$$

Therefore, the largest integer, $x + 2 = 101 + 2 = 103$.

Age Problems

Example 10: Phil is Tom's father. Phil is 35 years old. Three years ago, Phil was four times as old as his son was then. How old is Tom now?

First, circle what it is you must ultimately find—*how old is Tom now?* Therefore, let t be Tom's age now. Then three years ago, Tom's age would be $t - 3$. Four times Tom's age three years ago would be $4(t - 3)$. Phil's age three years ago would be $35 - 3 = 32$. A simple chart may also be helpful.

	now	*3 years ago*
Phil	35	32
Tom	t	$t - 3$

Now, use the problem to set up an equation.

Three years ago $\underbrace{\text{Phil}}_{32}$ $\underset{=}{\text{was}}$ $\underbrace{\text{four times}}_{4 \text{ times}}$ as old as his $\underbrace{\text{son was then.}}_{(t-3)}$

$$32 = 4(t-3)$$

$$\frac{32}{4} = \frac{4(t-3)}{4}$$

$$8 = t - 3$$

$$8 + 3 = t - 3 + 3$$

$$11 = t$$

Therefore, Tom is now 11.

Example 11: Lisa is 16 years younger than Kathy. If the sum of their ages is 30, how old is Lisa?

First, circle what you must find—*how old is Lisa?* Let Lisa equal x. Therefore, Kathy is $x + 16$. (Note that since Lisa is 16 years *younger* than Kathy, you must *add* 16 years to Lisa to denote Kathy's age.) Now, use the problem to set up an equation.

If $\underbrace{\text{the sum of their ages}}_{\text{Lisa's age + Kathy's age}}$ $\underset{=}{\text{is}}$ $\underbrace{30}_{30}$

$$x + (x + 16) = 30$$

$$2x + 16 = 30$$

$$2x + 16 - 16 = 30 - 16$$

$$2x = 14$$

$$\frac{2x}{2} = \frac{14}{2}$$

$$x = 7$$

Therefore, Lisa is 7 years old.

Motion Problems

Example 12: How long will it take a bus traveling 72 km/hr to go 36 kms?

First circle what you're trying to find—*how long will it take* (time). Motion problems are solved by using the equation

$$\text{distance = rate times time}$$

$$d = rt$$

Therefore, simply plug in: 72 km/hr is the rate (or speed) of the bus, and 36 km is the distance.

$$d = rt$$
$$36 \text{ km} = (72 \text{ km/hr})(t)$$
$$\frac{36}{72} = \frac{72t}{72}$$
$$\frac{1}{2} = t$$

Therefore, it will take one-half hour for the bus to travel 36 km at 72 km/hr.

Example 13: How fast in miles per hour must a car travel to go 600 miles in 15 hours?

First, circle what you must find—*how fast* (rate). Now, using the equation $d = rt$, simply plug in 600 for distance and 15 for time.

$$d = rt$$
$$600 = r(15)$$
$$\frac{600}{15} = \frac{r(15)}{15}$$
$$40 = r$$

So, the rate is 40 miles per hour.

Example 14: Mrs. Benevides leaves Burbank at 9 a.m. and drives west on the Ventura Freeway at an average speed of 50 miles per hour. Ms. Twill leaves Burbank at 9:30 a.m. and drives west on the Ventura Freeway at an average speed of 60 miles per hour. At what time will Ms. Twill overtake Mrs. Benevides, and how many miles will they each have gone?

First, circle what you are trying to find—*at what time and how many miles*. Now, let t stand for the time Ms. Twill drives before overtaking Mrs. Benevides. Then Mrs. Benevides drives for $t + \frac{1}{2}$ hours before being overtaken. Next, set up the following chart.

	rate r	×	time t	=	distance d
Ms. Twill	60 mph		t		$60t$
Mrs. Benevides	50 mph		$t + \frac{1}{2}$		$50\left(t + \frac{1}{2}\right)$

Because each travels the same distance,

$$60t = 50\left(t + \frac{1}{2}\right)$$

$$60t = 50t + 25$$

$$10t = 25$$

$$t = 2.5$$

Ms. Twill overtakes Mrs. Benevides after 2.5 hours of driving. The exact time can be figured out by using Ms. Twill's starting time: 9:30 + 2:30 = 12 noon. Since Ms. Twill has traveled for 2.5 hours at 60 mph, she has traveled 2.5 × 60, which is 150 miles. So, Mrs. Benevides is overtaken at 12 noon, and each has traveled 150 miles.

Coin Problems

Example 15: Tamar has four more quarters than dimes. If he has a total of $1.70, how many quarters and dimes does he have?

First, circle what you must find—*how many quarters and dimes.* Let x stand for the number of dimes, then $x + 4$ is the number of quarters. Therefore, .10x is the total value of the dimes, and .25($x + 4$) is the total value of the quarters. Setting up the following chart can be helpful.

	number	value	amount of money
dimes	x	.10	.10x
quarters	$x + 4$	.25	.25($x + 4$)

Now, use the table and problem to set up an equation.

$$.10x + .25(x + 4) = 1.70$$

$$10x + 25(x + 4) = 170$$

$$10x + 25x + 100 = 170$$

$$35x + 100 = 170$$

$$35x = 70$$

$$x = 2$$

So, there are two dimes. Since there are four more quarters, there must be six quarters.

Example 16: Sid has $4.85 in coins. If he has six more nickels than dimes and twice as many quarters as dimes, how many coins of each type does he have?

First, circle what you must find—*the number of coins of each type.* Let x stand for the number of dimes. Then $x + 6$ is the number of nickels, and $2x$ is the number of quarters. Setting up the following chart can be helpful.

number	value	amount of money	
dimes	x	.10	.10x
nickels	$x + 6$	.05	.05$(x + 6)$
quarters	$2x$	.25	.25$(2x)$

Now, use the table and problem to set up an equation.

$$.10x + .05(x + 6) + .25(2x) = 4.85$$

$$10x + 5(x + 6) + 25(2x) = 485$$

$$10x + 5x + 30 + 50x = 485$$

$$65x + 30 = 485$$

$$65x = 455$$

$$x = 7$$

So, there are seven dimes. Therefore, there are thirteen nickels and fourteen quarters.

Mixture Problems

Example 17: Coffee worth $1.05 per pound is mixed with coffee worth 85¢ per pound to obtain 20 pounds of a mixture worth 90¢ per pound. How many pounds of each type are used?

First, circle what you are trying to find—*how many pounds of each type.* Now, let the number of pounds of $1.05 coffee be denoted as x. Therefore, the number of pounds of 85¢-per-pound coffee must be the remainder of the twenty pounds, or $20 - x$. Now, make a chart for the cost of each type and the total cost.

	cost per lb. ×	amount in lbs. =	total cost of each
$1.05 coffee	$1.05	x	$1.05x$
$ .85 coffee	$ .85	20 − x	$ .85(20 − x)$
mixture	$ .90	20	$ .90(20)$

Now, set up the equation.

$$\underbrace{\text{total cost of one type}}_{\$1.05x} \underbrace{\text{plus}}_{+} \underbrace{\text{total cost of other type}}_{\$.85(20-x)} \underbrace{\text{equals}}_{=} \underbrace{\text{total cost of mixture}}_{\$.90(20)}$$

$$1.05x + 17.00 - .85x = 18.00$$
$$17.00 + .20x = 18.00$$
$$-17.00 + 17.00 + .20x = 18.00 - 17.00$$
$$.20x = 1.00$$
$$\frac{.20x}{.20} = \frac{1.00}{.20}$$
$$x = 5$$

Therefore, five pounds of coffee worth $1.05 per pound are used. And 20 − x, or 20 − 5, or fifteen pounds of 85¢-per-pound coffee are used.

Example 18: Solution A is 50% hydrochloric acid, while solution B is 75% hydrochloric acid. How many liters of each solution should be used to make 100 liters of a solution which is 60% hydrochloric acid?

First, circle what you're trying to find—*liters of solutions A and B*. Now, let x stand for the number of liters of solution A. Therefore, the number of liters of solution B must be the remainder of the 100 liters, or 100 − x. Next, make the following chart.

	% of acid	liters	concentration of acid
solution A	50%	x	$.50x$
solution B	75%	100 − x	$.75(100 − x)$
new solution	60%	100	$.60(100)$

Now, set up the equation.

$$.50x + .75(100 - x) = .60(100)$$
$$.50x + 75 - .75x = 60$$

$$\frac{\qquad -75 \qquad\qquad -75}{.50x \qquad -.75x = -15}$$

$$-.25x = -15$$
$$\frac{-.25x}{-.25} = \frac{-15}{-.25}$$
$$x = 60$$

Therefore, using the chart, 60 liters of solution A and 40 liters of solution B are used.

Work Problems

Example 19: Ernie can plow a field alone in four hours. It takes Sid five hours to plow the same field alone. If they work together (and each has a plow), how long will it take to plow the field?

First, circle what you must find—*how long . . . together.* Work problems of this nature may be solved by using the following equation.

$$\frac{1}{\text{1st person's rate}} + \frac{1}{\text{2nd person's rate}} + \frac{1}{\text{3rd person's rate}} + \text{etc.} = \frac{1}{\text{rate together}}$$

Therefore,

$$\frac{1}{\text{Ernie's rate}} + \frac{1}{\text{Sid's rate}} = \frac{1}{\text{rate together}}$$

$$\frac{1}{4} + \frac{1}{5} = \frac{1}{t}$$

Finding a common denominator

$$\frac{5}{20} + \frac{4}{20} = \frac{1}{t}$$

$$\frac{9}{20} = \frac{1}{t}$$

Cross multiplying

$$9t = 20$$

$$\frac{9t}{9} = \frac{20}{9} = 2\frac{2}{9} \text{ hours}$$

Therefore, it will take them $2\frac{2}{9}$ hours working together.

Number Problems with Two Variables

Example 20: The sum of two numbers is 15. The difference of the same two numbers is 7. What are the two numbers?

First, circle what you're looking for—*the two numbers.* Let x stand for the larger number and y stand for the second number. Now, set up two equations.

The sum of the two numbers is 15.

$$x + y = 15$$

The difference is 7.

$$x - y = 7$$

Now, solve by adding the two equations.

$$
\begin{aligned}
x + y &= 15 \\
\underline{x - y} &= \underline{7} \\
2x &= 22 \\
x &= 11
\end{aligned}
$$

Now, plugging into the first equation gives

$$11 + y = 15$$

So $\qquad\qquad\qquad y = 4$

The numbers are 11 and 4.

Example 21: The sum of twice one number and three times another number is 23 and their product is 20. Find the numbers.

First, circle what you must find—*the numbers.* Let x stand for the number that is being multiplied by 2 and y stand for the number being multiplied by 3.

Now set up two equations.

The sum of twice a number and three times another number is 23.

$$2x + 3y = 23$$

Their product is 20.

$$x(y) = 20$$

Rearranging the first equation gives

$$3y = 23 - 2x$$

Dividing each side of the equation by 3 gives

$$y = \frac{23}{3} - \frac{2x}{3}$$

Now, substituting the first equation into the second gives

$$x\left(\frac{23}{3} - \frac{2x}{3}\right) = 20$$

$$\frac{23x}{3} - \frac{2x^2}{3} = 20$$

Multiplying each side of the equation by 3 gives

$$23x - 2x^2 = 60$$

Rewriting this equation in standard quadratic form gives

$$2x^2 - 23x + 60 = 0$$

Solving this quadratic equation using factoring gives

$$(2x - 15)(x - 4) = 0$$

Setting each factor equal to 0 and solving gives

$$2x - 15 = 0 \text{ or } x - 4 = 0$$

$$2x = 15 \text{ or } \quad x = 4$$

$$x = \frac{15}{2} \text{ or } \quad x = 4$$

With each x value we can find its corresponding y value.

If $x = \frac{15}{2}$, then $y = \frac{23}{3} - \frac{2\left(\frac{15}{2}\right)}{3}$ or $y = \frac{23}{3} - \frac{15}{3} = \frac{8}{3}$.

If $x = 4$, then $y = \frac{23}{3} - \frac{2(4)}{3}$ or $y = \frac{23}{3} - \frac{8}{3} = \frac{15}{3} = 5$.

Therefore, this problem has two sets of solutions.

The number being multiplied by 2 is $\frac{15}{2}$, and the number being multiplied by 3 is $\frac{8}{3}$, or the number being multiplied by 2 is 4 and the number being multiplied by 3 is 5.

Chapter Check-Out

1. Find the total interest on $140 at 5% annual rate for two years if the interest is compounded annually.

2. A map's key shows that 1" = 50 miles. How many inches apart on the map will two cities be if they are exactly 15 miles apart?

3. Earl is six years older than Simin. In two years the sum of their ages will be twenty. How old is Simin now?

4. Ellen has collected nickels and dimes worth a total of $6.30. If she has collected seventy coins in all and each is worth face value, how many of each kind does she have?

5. Terrell can put up a wood fence in 5 hours if he works alone. It takes Miri 6 hours to put up the same wood fence if she works alone. If they work together, how long will it take them to put up the same wood fence?

6. The sum of two numbers is 40, and their product is 300. What are the two numbers?

Answers: 1. $14.35 **2.** $\frac{3"}{10}$ or .3" **3.** 5 **4.** 14 nickels and 56 dimes **5.** $2\frac{8}{11}$ hours **6.** 10 and 30

REVIEW QUESTIONS

Use these review questions to practice what you've learned in this book. After you work through the review questions, you're well on your way to understanding the basic concepts of Algebra I.

Chapter 1

1. Which of the following are rational numbers? 0, 2, $\sqrt{5}$, $\frac{1}{2}$, 0.6

2. Which of the following are prime numbers? 2, 6, 9, 11, 15, 17

3. The multiplicative inverse of $\frac{2}{3}$ is _____ .

4. $3^2 =$

5. $2^{-3} =$

6. $7^2 \times 7^5 =$ (with exponent)

7. $8^6 \div 8^4 =$ (with exponent)

8. $(3^4)^5 =$ (with exponent)

9. $\sqrt{64} =$

10. $\sqrt[3]{125} =$

11. Approximate $\sqrt{18}$ to the nearest tenth.

12. Simplify: $2[(3^2 + 4) + 2(1 + 2)]$.

13. The number 12,120 is divisible by which numbers between 1 and 10?

Chapter 2

14. $-7 + 6 =$

15. $(-4)(-2)(-3) =$

16. $\frac{3}{5} + \frac{2}{7} =$

17. $\frac{2}{3} \times \frac{15}{22} =$

18. $1\frac{1}{2} \times 2\frac{1}{4} =$

19. $3\frac{1}{3} \div 2\frac{1}{2} =$

20. Change to a decimal: $\frac{1}{16}$

21. Change to a fraction: $0.\overline{4}$

22. 20 is what % of 400?

For problems 23 and 24, express each answer in scientific notation.

23. $(4 \times 10^3)(3 \times 10^2) =$

24. $(8 \times 10^{-2}) \div (2 \times 10^4) =$

Chapter 3

25. $\{1, 2, 3\} \cap \{3, 4, 5\} =$

26. $\{1, 3, 5\} \cup \{3, 4, 5\} =$

27. True or false: $\{$Tom, Bob, Sam$\} \sim \{1, 2, 3\}$

28. Express algebraically: five less than four times a number n.

29. Evaluate: $2x + 4y^2$ if $x = 3$ and $y = -2$.

Chapter 4

30. Solve for x: $4x + 8 = 32$

31. Solve for y: $\frac{y}{8} - 3 = 9$

32. Solve for r: $5r + 7 = 3r - 15$

33. Solve for x: $xy + z = w$

34. Solve for c: $\frac{a}{c} = \frac{b}{d}$

35. Solve for x: $\frac{x}{4} = \frac{3}{9}$

Chapter 5

36. Solve for x and y:

$$3x + 2y = 1$$
$$2x - 3y = -8$$

37. Solve for m and n:

$$m = n + 3$$
$$m + 2n = 9$$

Chapter 6

38. $2xy^2 + xy^2 - 6xy^2 =$

39. $(3x^4y^2z)(-5x^2y^2z^2) =$

40. $(5a^2b)^4 =$

41. $\dfrac{9x^3y^6z^2}{3x^2y^5z}$

42. Express the answer with positive exponents: $x^{-5}yz^{-2}$

43. $(8a - 4b) - (6a - 3b) =$

44. $4x^2y - 3xy^2 + 2x^2y - 2xy^2 =$

45. $(5x + 2y)(2x + 3y) =$

46. $\dfrac{12m^2n^2 + 14mn^2}{2mn} =$

47. $(x^2 + 4x + 4) \div (x + 2) =$

48. Factor: $2y^2 - 8y =$

49. Factor: $9x^2 - 16 =$

50. Factor: $x^2 - 10x + 24 =$

51. Factor: $4a^3 + 6a^2 + 2a =$

52. Factor: $m^2 + 5mn + 4n^2 =$

53. Factor: $a + 6 + ab + 6b =$

Chapter 7

54. Reduce: $\dfrac{5x^6}{15x^4}$

55. Reduce: $\dfrac{x^2 - 3x + 2}{3x - 6}$

56. $\dfrac{x^3}{2y} \times \dfrac{3y^2}{4x} =$

57. $\dfrac{x^2 + 2x + 1}{x + 2} \times \dfrac{2x + 4}{x + 1}$

58. $\dfrac{6x^3}{7} \div \dfrac{2x^2}{y} =$

59. $\dfrac{6y + 12}{8} \div \dfrac{y + 2}{4} =$

60. $\dfrac{3x - 2}{x + 1} - \dfrac{2x - 1}{x + 1} =$

61. $\dfrac{5}{x} + \dfrac{7}{y} =$

62. $\dfrac{3}{a^3 b^3} + \dfrac{2}{a^4 b^2} =$

63. $\dfrac{3x}{x - 3} - \dfrac{2x}{x + 1} =$

64. $\dfrac{x}{x^2 - 16} + \dfrac{4x}{x^2 + 5x + 4} =$

Chapter 8

65. Solve for x: $2x + 5 < 15$

66. Solve for x: $3x + 4 \geq 5x - 8$

67. Graph: $\{x: -2 \leq x < 2\}$

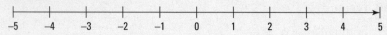

68. Graph: $\{x: x < 6\}$

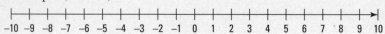

69. $|-6 - 3| =$

70. Solve for x: $2|x - 2| + 5 = 13$

71. Solve and graph: $3|x + 1| + 2 > 8$

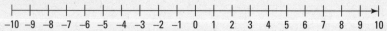

Chapter 9

72. Is $x + \dfrac{3}{y} = 9$ linear or nonlinear?

73. Graph: $y = x - 4$

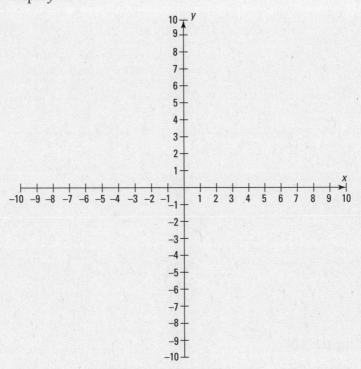

74. Find the slope of $2x + y = 6$.

75. Find the equation of the line, in slope-intercept form, passing through the point $(5, 3)$ with a slope of 2.

76. Find the equation of the line, in standard form, passing through the points $(2, 3)$ and $(-1, -3)$.

77. Graph: $y \geq x + 1$

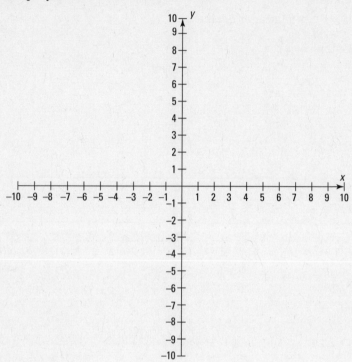

Chapter 10

78. Which of the following are graphs of functions?

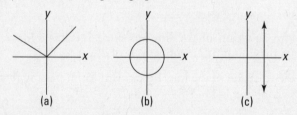

(a)   (b)   (c)

79. If $f(x) = 3x^2 - 2x - 1$, what is $f(3)$?

80. If the domain of $x^2 + 5x + 1$ is {2, 3, 4}, then what is the range?

81. If y varies directly as x, find the constant of variation when y is 3 and x is 9.

82. If y varies indirectly as x and the constant of variation is 2, find y when x is 8.

Chapter 11

83. Simplify: $\sqrt{75}$

84. If each variable is nonnegative, $\sqrt{25x^5y^8} =$

85. $\sqrt{50} + 3\sqrt{2} =$

86. $6\sqrt{3} \times 2\sqrt{2} =$

For problems 87–89, express answers in simplified form with rational denominators.

87. $\dfrac{\sqrt{7}}{\sqrt{2}} =$

88. $\dfrac{4}{2 - \sqrt{3}} =$

89. $\dfrac{3 + \sqrt{2}}{2 - \sqrt{2}} =$

Chapter 12

90. Solve for x: $x^2 - 2x = 63$

91. Solve: $x^2 - 81 = 0$

92. Solve: $x^2 + 8x = 0$

93. Solve: $3x^2 + 3x + 2 = 2x^2 + x + 1$

94. Solve for x using the quadratic formula: $x^2 + 3x + 1 = 0$

95. Solve for x by completing the square: $x^2 + 7x + 4 = 0$

Chapter 13

96. If Tim invests $200 at a 10% annual rate for three years compounded annually, how much money will he have at the end of three years from this investment?

97. If one number is twice as large as another number and the smaller number is increased by 12, the result is 8 less than the larger number. What is the larger number?

98. Nuts costing $1.40 per pound are mixed with nuts costing $1.00 per pound to produce forty pounds of mixture worth $1.10 per pound. How much of each type is used?

99. Tom is trying to fill his bathtub to take a bath. If Tom turns on only the hot water, the tub will fill in 20 minutes. If Tom turns on only the cold water, the tub will fill in 10 minutes. How long will it take to fill the tub if Tom turns on the hot and cold water faucets at the same time?

Answers: 1. $0, 2, \frac{1}{2}, 0.6$ **2.** $2, 11, 17$ **3.** $\frac{3}{2}$ or $1\frac{1}{2}$ **4.** 9 **5.** $\frac{1}{8}$ **6.** 7^7 **7.** 8^2 **8.** 3^{20}

9. 8 **10.** 5 **11.** approx. 4.2 **12.** 38 **13.** $2, 3, 4, 5, 6, 8$ **14.** -1 **15.** -24

16. $\frac{31}{35}$ **17.** $\frac{5}{11}$ **18.** $\frac{27}{8}$ or $3\frac{3}{8}$ **19.** $\frac{4}{3}$ or $1\frac{1}{3}$ **20.** $.0625$ **21.** $\frac{4}{9}$ **22.** 5%

23. $12 \times 10^5 = 1.2 \times 10^6$ **24.** 4×10^{-6} **25.** $\{3\}$ **26.** $\{1, 3, 4, 5\}$ **27.** True

28. $4n - 5$ **29.** 22 **30.** 6 **31.** 96 **32.** -11 **33.** $\frac{w - z}{y}$ **34.** $\frac{ad}{b}$ **35.** $\frac{12}{9} = \frac{4}{3} = 1\frac{1}{3}$

36. $x = -1, y = 2$ **37.** $m = 5, n = 2$ **38.** $-3xy^2$ **39.** $-15x^6y^4z^3$ **40.** $625a^8b^4$

41. $3xyz$ **42.** $\frac{y}{x^5z^2}$ **43.** $2a - b$ **44.** $6x^2y - 5xy^2$ **45.** $10x^2 + 19xy + 6y^2$

46. $6mn + 7n$ **47.** $x + 2$ **48.** $2y(y - 4)$ **49.** $(3x - 4)(3x + 4)$ **50.** $(x - 6)(x - 4)$

51. $2a(2a + 1)(a + 1)$ **52.** $(m + n)(m + 4n)$ **53.** $(1 + b)(a + 6)$ **54.** $\frac{x^2}{3}$ **55.** $\frac{x - 1}{3}$

56. $\frac{3x^2y}{8}$ **57.** $2(x + 1)$ **58.** $\frac{3xy}{7}$ **59.** 3 **60.** $\frac{x - 1}{x + 1}$ **61.** $\frac{5y + 7x}{xy}$ or $\frac{7x + 5y}{xy}$

62. $\frac{3a + 2b}{a^4b^3}$ **63.** $\frac{x(x + 9)}{(x - 3)(x + 1)}$ or $\frac{x^2 + 9x}{x^2 - 2x - 3}$ **64.** $\frac{5x^2 - 15x}{(x + 4)(x - 4)(x + 1)}$

65. $x < 5$ **66.** $6 \geq x$ or $x \leq 6$

67.

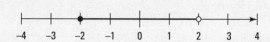

68.

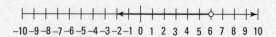

69. 9 **70.** 6, – 2 **71.** $x > 1$ or $x < -3$

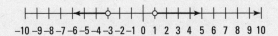

72. Nonlinear

73.

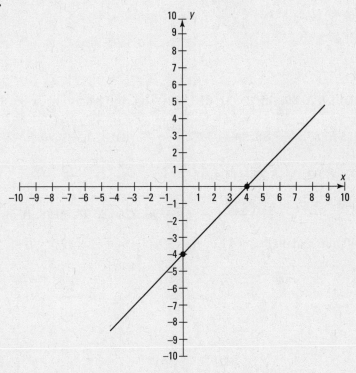

74. –2 **75.** $y = 2x - 7$ **76.** $2x - y = 1$

77.

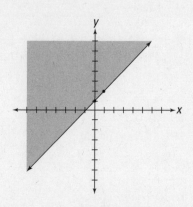

78. (a) **79.** 20 **80.** {15, 25, 37} **81.** $\frac{1}{3}$ **82.** $\frac{1}{4}$ **83.** $5\sqrt{3}$ **84.** $5x^2y^4\sqrt{x}$ **85.** $8\sqrt{2}$

86. $12\sqrt{6}$ **87.** $\frac{\sqrt{14}}{2}$ **88.** $8+4\sqrt{3}$ **89.** $\frac{8+5\sqrt{2}}{2}$ or $4+\frac{5\sqrt{2}}{2}$ **90.** 9, −7

91. 9, −9 **92.** 0, −8 **93.** −1 **94.** $\frac{-3\pm\sqrt{5}}{2}$ or $\frac{-3+\sqrt{5}}{2}$, $\frac{-3-\sqrt{5}}{2}$

95. $\frac{-7\pm\sqrt{33}}{2}$ or $\frac{-7+\sqrt{33}}{2}$, $\frac{-7-\sqrt{33}}{2}$ **96.** $266.20 **97.** 40 **98.** 10 lbs. of

$1.40 nuts and 30 lbs. of $1.00 nuts **99.** $6\frac{2}{3}$ min.

RESOURCE CENTER

The Resource Center offers the best resources available in print and online to help you study and review the core concepts of algebra. You can find additional resources, plus study tips and tools to help test your knowledge, at www.cliffsnotes.com.

Books

CliffsNotes Algebra I Quick Review, 2nd Edition, is one of many great books available to help you review, refresh, and relearn mathematics. If you want some additional resources for math review, check out the following publications.

CliffsNotes Math Review for Standardized Tests, 2nd Edition, by Jerry Bobrow, Ph.D., gives you a step-by-step review of arithmetic, algebra, geometry, and word problems. Includes pre-test and post-test for each subject area and a glossary. Published by Wiley $14.99.

CliffsNotes Basic Math and Pre-Algebra Quick Review, 2nd Edition, by Jerry Bobrow, Ph.D., gives you an easy-to-use guide to review arithmetic and prepare for Algebra I. Includes chapter checkouts, chapter reviews, and a glossary. Published by Wiley $9.99.

CliffsQuickReview Geometry by Edward Kohn, M.S., gives you an outstanding review of the basic concepts of geometry. Includes chapter checkouts, chapter reviews, and a glossary. Published by Wiley $9.99.

CliffsNotes Algebra II Quick Review, 2nd Edition, by Edward Kohn, M.S., and David Alan Herzog, provides an excellent review of Algebra II with clearly explained sample problems. Includes chapter checkouts, chapter reviews, and a glossary. Published by Wiley $9.99.

CliffsQuickReview Trigonometry by David A. Kay, M.S., provides an outstanding review of the basic concepts of trigonometry. Clearly explained example problems and figures make the concepts easier to understand. Includes chapter checkouts, chapter reviews, and a glossary. Published by Wiley $9.99.

Wiley also has three Web sites that you can visit to read about all the books we publish:

■ www.cliffsnotes.com

■ www.dummies.com

■ www.wiley.com

Internet

Visit the following Web sites for more information about algebra.

Math.com, http://www.math.com/students/homeworkhelp.html. If you can spare a mere 60 seconds, then you can improve your math skills. Math.com's Homework Help offers quick one-minute tips, tricks, and tidbits designed to help you improve your skills in Pre-Algebra, Algebra, and Geometry. Math.com also offers a handy glossary that helps you figure out the difference between a rhombus and a ray.

Discovery.com, http://school.discovery.com/homeworkhelp/webmath/. Have you been working on the same math problem for an hour? Help is on the way! At Discovery.com's WebMath you can find tips and tricks on every type of math problem from Pre-Algebra to Calculus. Just find a math problem similar to the one you're working on, then use WebMath to find out how to solve the problem and check your answer. Your math homework just got a whole lot easier.

Next time you're on the Internet, don't forget to drop by www.cliffs notes.com. We created an online Resource Center that you can use today, tomorrow, and beyond.

GLOSSARY

abscissa the distance along the horizontal axis in a coordinate graph.

absolute value the numerical value when direction or sign is not considered. The symbol for absolute value is $|\ |$.

additive axiom of equality if $a = b$ and $c = d$, then $a + c = b + d$.

additive axiom of inequality if $a > b$, then $a + c > b + c$.

additive inverse the opposite (negative) of a number. Any number plus its additive inverse equals 0.

algebra arithmetic operations using letters and/or symbols in place of numbers.

algebraic expressions expressions composed of letters to stand for numbers.

algebraic fractions fractions using a variable in the numerator and/or denominator.

ascending order basically, when the power of a term increases for each succeeding term.

associative property grouping of elements does not make any difference in the outcome. Only true for multiplication and addition.

axioms of equality basic rules for using the equal sign.

binomial an algebraic expression consisting of two terms.

braces grouping symbols used after the use of brackets. Also used to represent a set. { }

brackets grouping symbols used after the use of parentheses. [ ]

canceling in multiplication of fractions, dividing the same number into both a numerator and a denominator.

Cartesian coordinates a system of assigning ordered number pairs to points on a plane.

closed half-plane a half-plane that includes the boundary line and is graphed using a solid line and shading.

closed interval an interval that includes both endpoints or fixed boundaries.

closure property when all answers fall into the original set.

coefficient a number or variable affixed to another by multiplication. For example, in $9x$, 9 is the numerical coefficient of x and x is the variable coefficient of 9.

common factors factors that are the same for two or more numbers.

commutative property order of elements does not make any difference in the outcome. Only true for multiplication and addition.

complex fraction a fraction having a fraction or fractions in the numerator and/or denominator.

composite number a number divisible by more than just 1 and itself (such as 4, 6, 8, 9, . . .). 0 and 1are not composite numbers.

conjugate the conjugate of a binomial contains the same terms, but the opposite sign between them. $(x + y)$ and $(x - y)$ are conjugates.

coordinate axes two perpendicular number lines used in a coordinate graph.

coordinate graph two perpendicular number lines, the x axis and the y axis, creating a plane on which each point is assigned a pair of numbers.

coordinates the numbers that correspond to a point on a coordinate graph.

cube the result when a number is multiplied by itself twice. Designated by the exponent 3 (such as x^3).

cube root the number that when multiplied by itself twice gives you the original number. For example, 5 is the cube root of 125.

denominator everything below the fraction bar in a fraction.

descending order basically, when the power of a term decreases for each succeeding term.

direct variation when y varies directly as x or y is directly proportional to x.

discriminant the value under the radical sign in the quadratic formula. $[b^2 - 4ac]$

distributive property the process of distributing the number on the outside of the parentheses to each number on the inside. $a(b + c) = ab + ac$

domain the set of all first coordinates from the ordered pairs in a relation.

element a member of a set.

empty set a set with no members (a null set).

equal sets sets that have exactly the same members.

equation a balanced relationship between numbers and/or symbols. A mathematical sentence.

equivalent sets sets that have the same number of members.

Euler circles a method of pictorially representing sets.

evaluate to determine the value or numerical amount.

exponent a numeral used to indicate the power of a number.

extremes outer terms.

factor to find two or more quantities whose product equals the original quantity.

finite countable. Having a definite ending.

F.O.I.L. method a method of multiplying binomials in which first terms, outside terms, inside terms, and last terms are multiplied.

function a relation in which each element in the domain is paired with exactly one element in the range.

graphing method a method of solving simultaneous equations by graphing each equation on a coordinate graph and finding the common point (intersection).

half-open interval an interval that includes one endpoint, or one boundary.

half-plane the region of a coordinate graph on one side of a boundary line.

identity element for addition 0. Any number added to 0 gives the original number.

identity element for multiplication 1. Any number multiplied by 1 gives the original number.

imaginary numbers square roots of negative numbers. The imaginary unit is i.

incomplete quadratic equation a quadratic equation with a term missing.

indirect variation or inverse variation when y varies indirectly as x or y is indirectly proportional to x. That is, as x increases, y decreases and as y increases, x decreases. Also referred to as inverse or indirect proportion.

inequality a statement in which the relationships are not equal. The opposite of an equation.

infinite uncountable. Continues forever.

integer a whole number, either positive, negative, or zero.

intersection of sets the members that overlap (are in both sets).

interval all the numbers that lie within two certain boundaries.

inverse relations relations where the domain and the range have been interchanged—switching the coordinates in each ordered pair.

linear equation an equation whose solution set forms a straight line when plotted on a coordinate graph.

literal equation an equation having mostly variables.

means inner terms in a proportion.

monomial an algebraic expression consisting of only one term.

multiplicative axiom of equality if $a = b$ and $c = d$, then $ac = bd$.

multiplicative inverse the reciprocal of the number. Any number multiplied by its multiplicative inverse equals 1.

negative multiplication property of inequality reverse the inequality sign when multiplying (or dividing) by a negative number. If $c < 0$, then $a > b$ if, and only if, $ac < bc$.

nonlinear equation an equation whose solution set does not form a straight line when plotted on a coordinate graph.

null set a set with no members (an empty set).

number line a graphic representation of integers and real numbers. The point on this line associated with each number is called the graph of the number.

numerator everything above the fraction bar in a fraction.

numerical coefficient the number in front of the variable.

open half-plane a half-plane that does not include the boundary line. If the inequality is a ">" or "<", then the graph is an open half-plane.

open interval an interval that does not include endpoints or fixed boundaries.

open ray a ray that does include its endpoint (half line).

ordered pair any pair of elements (x, y) having a first element x and a second element y. Used to identify or plot points on a coordinate grid.

ordinate the distance along the vertical axis on a coordinate graph.

origin the point of intersection of the two number lines on a coordinate graph. Represented by the coordinates $(0, 0)$.

polynomial an algebraic expression consisting of two or more terms.

positive multiplication property of inequality if $c > 0$, then $a > b$ if, and only if, $ac > bc$.

proportion two ratios equal to each other. For example, a is to c as b is to d.

quadrants four quarters or divisions of a coordinate graph.

quadratic equation an equation that could be written $ax^2 + bx + c = 0$.

quadratic formula a method of solving quadratic equations.

radical sign the symbol used to designate square root.

range the set of all second (or y) coordinates from the ordered pairs in a relation.

ratio a method of comparing two or more numbers. For example, $a:b$. Often written as a fraction, a/b.

real numbers the set consisting of all rational and irrational numbers.

reducing changing a numerical or algebraic fraction into its lowest terms. For example, $2/4$ is reduced to $1/2$, or a/ab is reduced to $1/b$.

reflexive axiom of equality for any number a, $a = a$.

relation any set of ordered pairs.

repeating decimal a decimal fraction that continues forever repeating a number or block of numbers.

roster a method of naming a set by listing its members.

rule a method of naming a set by describing its elements.

set a group of objects, numbers, and so forth.

set builder notation a formal method of describing a set. Often used for inequalities. For example, $\{x: x > 1\}$, which is read "x such that all x is greater than 1."

simplify to combine several or many terms into fewer terms.

simultaneous equations (system of equations) a set of equations with the same unknowns (variables).

slope of a line the ratio of the change in y to the change in x in a linear equation (slope = rise/run).

solution set (or solution) all the answers that satisfy the equation.

square the result when a number is multiplied by itself. Designated by the exponent 2 (such as x^2).

square root the number that when multiplied by itself gives you the original number. For example, 5 is the square root of 25.

subset   a set within a set.

substitution method   a method of solving simultaneous equations that involves substituting one equation into another.

symmetric axiom of equality   if $a = b$, then $b = a$.

system of equations   simultaneous equations.

term   a numerical or literal expression with its own sign.

transitive axiom of equality   if $a = b$ and $b = c$, then $a = c$.

transitive axiom of inequality   if $a > b$ and $b > c$, then $a > c$. Or if $a < b$ and $b < c$, then $a < c$.

trichotomy axiom of inequality   the only possible relationships between two numbers are: $a > b$, $a = b$, or $a < b$.

trinomial   an algebraic expression consisting of three terms.

union of sets   all the numbers in those sets.

universal set   the general category set, or the set of all those elements under consideration.

unknown   a letter or symbol whose value is not known, a variable.

value   numerical amount.

variable   a symbol used to stand for a number.

variation   a relationship between a set of values of one variable and a set of values of other variables.

Venn diagram   a pictorial description of sets.

vinculum   a line placed over (sometimes under) a digit or group of digits in a repeating decimal fraction to show which digits are repeating.

whole number   0, 1, 2, 3, and so on.

x-axis   the horizontal axis in a coordinate graph.

x-coordinate   the first number in the ordered pair. Refers to the distance on the x-axis (the abscissa).

y-axis   the vertical axis in a coordinate graph.

y-coordinate   the second number in the ordered pair. Refers to the distance on the y-axis (the ordinate).

Index

SYMBOLS